Climate Change

American and Comparative Environmental Policy

Sheldon Kamieniecki and Michael E. Kraft, series editors

Russell J. Dalton, Paula Garb, Nicholas P. Lovrich, John C. Pierce, and John M. Whiteley, Critical Masses: Citizens, Nuclear Weapons Production, and Environmental Destruction in the United States and Russia

Daniel A. Mazmanian and Michael E. Kraft, eds., Toward Sustainable Communities: Transition and Transformations in Environmental Policy

Elizabeth R. DeSombre, Domestic Sources of International Environmental Policy: Industry, Environmentalists, and U.S. Power

Kate O'Neill, Waste Trading among Rich Nations: Building a New Theory of Environmental Regulation

Joachim Blatter and Helen Ingram, eds., Reflections on Water: New Approaches to Transboundary Conflicts and Cooperation

Paul F. Steinberg, Environmental Leadership in Developing Countries: Transnational Relations and Biodiversity Policy in Costa Rica and Bolivia

Uday Desai, ed., Environmental Politics and Policy in Industrialized Countries

Kent Portney, Taking Sustainable Cities Seriously: Economic Development, the Environment, and Quality of Life in American Cities

Edward P. Weber, Bringing Society Back In: Grassroots Ecosystem Management, Accountability, and Sustainable Communities

Norman J. Vig and Michael G. Faure, eds., Green Giants? Environmental Policies of the United States and the European Union

Robert F. Durant, Daniel J. Fiorino, and Rosemary O'Leary, eds., Environmental Governance Reconsidered: Challenges, Choices, and Opportunities

Paul A. Sabatier, Will Focht, Mark Lubell, Zev Trachtenberg, Arnold Vedlitz, and Marty Matlock, eds., Swimming Upstream: Collaborative Approaches to Watershed Management

Sally K. Fairfax, Lauren Gwin, Mary Ann King, Leigh S. Raymond, and Laura Watt, Buying Nature: The Limits of Land Acquisition as a Conservation Strategy, 1780–2004

Steven Cohen, Sheldon Kamieniecki, and Matthew A. Cahn, Strategic Planning in Environmental Regulation: A Policy Approach That Works

Michael E. Kraft and Sheldon Kamieniecki, eds., Business and Environmental Policy: Corporate Interests in the American Political System

Joseph F. C. DiMento and Pamela Doughman, eds., Climate Change: What It Means for Us, Our Children, and Our Grandchildren

Climate Change

What It Means for Us, Our Children, and Our Grandchildren

Edited by Joseph F. C. DiMento and
Pamela Doughman

The MIT Press
Cambridge, Massachusetts
London, England

For information about special quantity discounts, please email special_sales@mitpress.mit.edu

This book was set in Sabon on 3B2 by Asco Typesetters, Hong Kong. Printed on recycled paper and bound in the United States of America.

Library of Congress Cataloging-in-Publication Data

Climate change : what it means for us, our children, and our grandchildren / edited by Joseph F. C. DiMento, Pamela Doughman.
p. cm.—(American and comparative environmental policy)
Includes bibliographical references and index.
ISBN 978-0-262-04241-3 (alk. paper)—ISBN 978-0-262-54193-0 (pbk. : alk. paper)
1. Climatic changes—History. 2. Climatic changes—Social aspects. 3. Communication in science. I. DiMento, Joseph F. II. Doughman, Pamela.
QC981.8.C5C6126 2007
577.2′2—dc22 2006035345

Pamela Doughman is an Energy Specialist with the California Energy Commission. Her contributions to this book were made in an individual capacity and not as an employee of the California Energy Commission. The opinions, conclusions, and findings expressed in her work in this book are hers alone, and do not necessarily express the official position, policies, or opinions of the California Energy Commission or the State of California.

10 9 8 7 6 5 4 3 2 1

Contents

Series Foreword

On Memorial Day weekend in 2004, 20th Century Fox studios released a $125 million science fiction film, *The Day after Tomorrow*. The film grossly exaggerates the effects of climate change on the planet by showing scenes of tornadoes in Los Angeles and massive floods in New York City. By the end of the movie, the entire planet has succumbed to an ice age. The movie grossed nearly half a billion dollars worldwide within one month after its release. Despite the fantasy nature of the movie, the media coverage that it generated drew the public's attention to the climate-change issue. In fact, surveys conducted before and after the release of the film indicated that it significantly affected the climate-change risk perceptions, conceptual frameworks, behavioral intentions, policy positions, and voting intentions of those who saw the movie. Anecdotal evidence suggests that the 2006 documentary film *An Inconvenient Truth*, produced by and featuring former Vice President Al Gore, has also increased public concerns about climate change.

The debate over climate change involves a broad range of interest groups that have concerns that sometimes overlap but frequently diverge. Environmental groups have spent substantial time and resources on trying to persuade political leaders to curtail greenhouse-gas (GHG) emissions. Even a moderately aggressive program to curtail greenhouse-gas emissions is expensive, and its high costs will be paid by both fossil-fuel producers and fossil-fuel consumers, including average citizens and a wide variety of small and large companies. Insurance companies are concerned about significant claims related to increasingly severe weather. For all these reasons, scientists are under pressure to produce research findings that clearly point to what the causes of the problem are, what its scope and severity are, how much must be done, and how quickly action must be taken. Additional research on climate change will require significant government funding over a considerable period of time because of the complexity of the issue. Meanwhile, American politicians and policymakers are caught in the middle of a contentious debate and must keep a vigilant eye on the results of scientific studies and public opinion. Unless the danger from climate change is perceived to be real and immediate action is warranted, the public is likely to reject required and costly lifestyle changes.

This book introduces readers to the science, politics, and policies of climate change. Its main goal is to educate students and members of the general public about the scientific and political issues concerning climate change by providing balanced and well-documented information and observations about the problem. It notes the efforts that have been made by fossil-fuel producers and consumers to distort the discourse over climate change and persuade the public that a problem does not exist. The book assumes three things—that the public wants to know more about climate change, that understanding climate change is not easy, and that it does not have to be that way. The study analyzes the disconnection between the scientific community's assessment of the importance of the problem and the inaction of some governments, principally the U.S. national government. Contributors to this book provide a clear, unbiased, and coherent presentation of the various elements associated with the debate over this significant issue.

The book begins with a primer on global climate change and on its effects on the world, on certain regions, and on individual nations. It then reports on the current level of scientific understanding of why the earth is warming and what exactly is causing this. Following these science-oriented chapters, other contributors explore global responses in the public and private sectors. The media have played a key role in communicating to the public information about this issue, and one eminent journalist suggests ways in which the media can do a better job of educating the public about this problem. This is followed by an assessment of the impact of climate change on developing countries and on a vulnerable city in the United States. The last chapter pushes readers to address the climate-change problem in light of existing scientific knowledge. As governments postpone action, the climate-change problem becomes worse and more costly to correct. Concerned citizens and students will come away from this book with a basic understanding of the climate-change controversy.

The analyses presented in this book illustrate our purpose in the MIT Press series on American and Comparative Environmental Policy. We encourage work that examines a broad range of environmental policy issues. We are particularly interested in volumes that incorporate interdisciplinary research and focus on the linkages between public policy and environmental problems and issues both within the United States and in cross-national settings. We welcome contributions that analyze the policy dimensions of relationships between humans and the environment from either a theoretical or empirical perspective. At a time when environmental policies are increasingly seen as controversial and new approaches are being implemented widely, we especially encourage studies that assess policy successes and failures, evaluate new institutional arrangements and policy tools, and clarify new directions for environmental politics and policy. The books in this series are written for a wide audience that includes academics, policymakers, environmental scientists and professionals, business and labor leaders, environmental activists, and students concerned with environmental issues. We hope they contribute to public understanding of environmental problems, issues, and policies of concern today and also suggest promising actions for the future.

Sheldon Kamieniecki, *University of California, Santa Cruz*
Michael E. Kraft, *University of Wisconsin, Green Bay*
American and Comparative Environmental Policy Series Editors

Acknowledgments

This work came about from the collective contributions of a number of people to whom we owe significant thanks. The original idea for the book evolved from a program of the Newkirk Center for Science and Society at the University of California, Irvine (UCI). The Center supported book production throughout its many stages. The Center is generously funded by a major gift from Martha and James Newkirk who also supplied supplemental funding specifically for this work.

We are responsible for the analysis, but it is a more complete treatment thanks to the serious, substantive, and detailed comments that the anonymous MIT reviewers supplied.

Additional funding for the program and support of the Center came from UCI's Division of Research and Graduate Studies, the School of Social Ecology; the Research Group in International Environmental Cooperation, the Center for Global Peace and Conflict Studies; the Canadian Consulate, Los Angeles; the University of California Institute for Global Conflict and Cooperation; and the California Climate Change Registry.

To a large number of scientists and analysts we are particularly indebted for comments, advice and corrections. Most notable are professors Susan Trumbore and Eric Salzman, Earth System Science, UCI, and James Fleming, Colby College.

Clay Morgan and Meagan Stacey of the MIT Press made it a pleasure to produce this work and showed wonderful patience with the logistics of an edited volume with several authors. Sheldon Kamieniecki and Michael E. Kraft, editors of the MIT Press American and Comparative Environmental Policy series, were supportive through the review process and provided helpful insights. The editing skills of Deborah Cantor-Adams and Rosemary Winfield helped us to realize our goal of clearly communicating a complex subject. Production of *Climate Change* was kept on time and professional through the extra efforts of Marlene Dyce, Elizabeth Eastin, T. J. Fudge, and Shyla Raghav of the Newkirk Center for Science and Society.

We are grateful to the American Association for the Advancement of Science and the History of Science Society for supporting the George Sarton Memorial Lecture that led to Professor Oreskes's chapter, to Myles Allen and David Stainforth for permission to reproduce the figure from climateprediction.net, and to Erik Conway, Alison MacFarlane, and Leonard Smith for comments on early versions of that chapter.

Climate Change

1

Introduction: Making Climate Change Understandable

Joseph F. C. DiMento and Pamela Doughman

Over the centuries, humans have tried to change the weather. People have prayed, danced, seeded clouds, and used other strategies to get more rain, to stop the rain, to decrease the heat, and to warm things up a bit. Seldom have we deliberately tried to change *climate*—the average weather conditions over an extended period of time—but we have unintentionally changed climate historically and we are changing it today.

This book draws on the vast knowledge of earth-system science to explore changes in climate, including important changes linked to the level of what are known as *greenhouse gases*—the 3 percent of the gases in the earth's atmosphere that help to warm the planet. It addresses how such changes may affect us, our children, and our grandchildren—globally and locally.

Global climate change is a major societal issue that many citizens do not understand, do not take seriously, and do not consider to be a major public-policy concern. At most, as Bill McKibben notes in *Granta*, they think of climate change as they do of the trade deficit, violence, and television—"as a marginal concern for them . . . if a concern at all" (McKibben 2003). Californians are a bit of an exception, although their understanding probably is not much different than that of other Americans.[1] Yet the scientific community, with the exception of a few contrarians, sees climate as one of the major challenges facing society in the next decades. This book aims to make climate change understandable to the educated public.

We have a number of friends who respond to assertions that climate change is a problem by saying that concerns are part of "a disaster strategy." They feel that scientists and policymakers articulate dire environmental futures because it is in their professional interests to do so. One developer friend put it this way:

> I remember years ago during a short downturn in the fish catch in Upper Newport Bay, California, we were told that the situation was hopeless and that the future was one of us being fished out. This year, like others back and forth, we have had great catches. I just don't believe some of these scare scenarios.[2]

The scientific community's assessment of the importance of climate change has not persuaded some governments to take actions to address the problem. While many environmental risks (such as nuclear contamination) arouse greater anxiety among policymakers and the public than among members of the scientific community, climate change seems to produce the reverse result, at least in the United States.

Not everyone downplays the threat of global climate change, however. Internationally and nationally, industry and some governments are responding to the science of climate change in ways that will affect us and our grandchildren.

This book assumes three things: that the public would like to understand climate change better, that understanding climate change is not easy, and that it doesn't have to be that way.

Obstacles to Understanding Climate Change

Climate change can be difficult to understand for four key reasons.

Some people believe that scientists lack consensus on the human contribution to climate change.

In part, a public perception of lack of consensus has been powered by the strong statements made by politicians who have selectively parsed the words of mainstream scientists and used the conclusions of those who are outliers on the subject. One outspoken United States senator regularly refers to global climate change as the largest hoax ever perpetrated on the American people. In part, the perception of a weak scientific base comes from fictional works like Michael Crichton's novel. To write a compelling piece of fiction, he uses footnotes and other gimmicks to give the impression that science is being abused by climate-change investigators.

Governments also use ambiguous language to promote particular policy positions. A White House staff member, for instance, recently edited governmental climate reports in ways that could have major implications for the public's understanding of the seriousness of the climate issue (figure 1.1). Adding a phrase like "significant and fundamental" before "uncertainties" can "tend to produce an air of doubt about findings that most climate experts say are robust" (Revkin 2005).

Amplifying uncertainties about climate change is consistent with the national politics of the United States and mirrors the minimizing of uncertainties that some environmental groups have done. Outcomes that are "likely" can easily convert to statements of either fact or uncertainty.

The public's conclusion that scientific consensus on climate change has not been reached also derives from the way climate change has been treated in the communication media. Ethical journalists are committed to presenting controversial subjects fairly and to searching out contrarian view points. But readers of newspaper and magazine articles and viewers of television news programs are left with an impression that

An Editor in the White House

Handwritten revisions and comments by Philip A. Cooney, chief of staff for the White House Council on Environmental Quality, appear on two draft reports by the Climate Change Science Program and the Subcommittee on Global Change Research. Mr. Cooney's changes were incorporated into later versions of each document, shown below with revisions in bold.

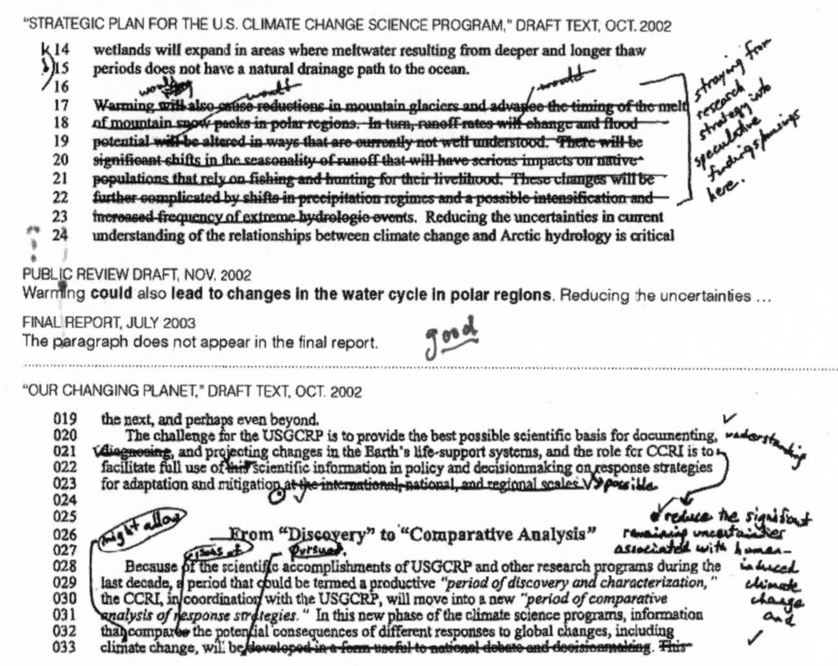

"STRATEGIC PLAN FOR THE U.S. CLIMATE CHANGE SCIENCE PROGRAM," DRAFT TEXT, OCT. 2002

14 wetlands will expand in areas where meltwater resulting from deeper and longer thaw
15 periods does not have a natural drainage path to the ocean.
16
17 Warming will also cause reductions in mountain glaciers and advance the timing of the melt
18 of mountain snow packs in polar regions. In turn, runoff rates will change and flood
19 potential will be altered in ways that are currently not well understood. There will be
20 significant shifts in the seasonality of runoff that will have serious impacts on native
21 populations that rely on fishing and hunting for their livelihood. These changes will be
22 further complicated by shifts in precipitation regimes and a possible intensification and
23 increased frequency of extreme hydrologic events. Reducing the uncertainties in current
24 understanding of the relationships between climate change and Arctic hydrology is critical

PUBLIC REVIEW DRAFT, NOV. 2002

Warming **could** also **lead to changes in the water cycle in polar regions**. Reducing the uncertainties ...

FINAL REPORT, JULY 2003

The paragraph does not appear in the final report.

"OUR CHANGING PLANET," DRAFT TEXT, OCT. 2002

019 the next, and perhaps even beyond.
020 The challenge for the USGCRP is to provide the best possible scientific basis for documenting,
021 diagnosing, and projecting changes in the Earth's life-support systems, and the role for CCRI is to
022 facilitate full use of this scientific information in policy and decisionmaking on response strategies
023 for adaptation and mitigation at the international, national, and regional scales.
024
025
026 From "Discovery" to "Comparative Analysis"
027
028 Because of the scientific accomplishments of USGCRP and other research programs during the
029 last decade, a period that could be termed a productive *"period of discovery and characterization,"*
030 the CCRI, in coordination with the USGCRP, will move into a new *"period of comparative*
031 *analysis of response strategies."* In this new phase of the climate science programs, information
032 that compares the potential consequences of different responses to global changes, including
033 climate change, will be developed in a form useful to national debate and decisionmaking. This

FINAL REPORT, 2003

The challenge for the USGCRP is to provide the best possible scientific basis for documenting, **understanding**, and projecting changes in the Earth's life-support systems, and the role for CCRI is to **reduce the significant remaining uncertainties associated with human-induced climate change and** facilitate full use of scientific information in policy and decisionmaking on **possible** response strategies for adaptation and mitigation.

Figure 1.1
An edited climate report

climate change is both a major worry and nothing to be concerned about.

The media also tend not to provide in-depth coverage of environmental issues—especially one that does not have immediate dramatic effects. Responsible ongoing coverage will seldom impress a public bombarded with more easily graspable stories. Occasionally, climate change is a front-page news item, as with the case of hurricane damage, but later coverage might question whether the original story was merited or even accurate. When accurate science stories reach the popular media's front pages or major segments, they often are packaged in a way that seems exaggerated.

In part, a perception of nonconsensus on climate change results from the nature of the science, which is based both on computer models and also on actual tests, measurements, and completed studies in the field. Computer models are powerful tools that make assumptions that are not always considered credible by critics, and even mainline scientists agree that the models need refinement. For example, one of our authors has written that the models need to reflect "improving understanding of the aerosols spewed by smokestacks, unfiltered tailpipes and volcanoes. They were once presumed only to have a cooling influence. Now, however, aerosols are known to cause both cooling and warming, depending on their color and composition and how they affect clouds, whose properties are slowly being incorporated in the simulations" (Revkin 2004).

Based on a growing body of observations, field research, ice-core drilling, and increasingly detailed computer models, a majority of scientists share the opinion that climate change is real, serious, and to an important extent, human induced. Clearly measuring and communicating how much, within convincing and influential ranges, is human caused and how much results from "natural variability" would be an important step in educating the public and perhaps influencing policymakers.

Scientists work with probabilities, risks, ranges, uncertainties, and "scenarios"—approaches that are foreign to many citizens. People often learn of scientific findings from experts who are not trained in communication or are trained to communicate only with their peers. Furthermore, models that earth-system scientists, atmospheric chemists, and others consider simple are not easy to follow for the nonscientifically educated person.

The vocabulary, science, and policies of climate change are complex.

The climate field is peppered with terms such as *sinks*, *forcing*, *secondary effects*, *adaptive capacity*, *albedo*, *carbon cycle*, *integrated assessment*, *no-regrets policy*, *net primary production*, *joint development*, and *clean-development mechanism* and many acronyms. In fact, there are 288 terms in the glossary of terms of one assessment report of one working group of the Intergovernmental Panel on Climate Change (IPCC). The IPCC, created in 1988, is the main international body established by the World Meteorological Organization and the United Nations Environment Program to assess climate-change science and provide advice to the international community. In this book, we use everyday language terms or define a term that is not common when it is first used. We use the terms *anthropogenic* and *human* interchangeably.

The environmental and social effects of climate change are not discreet.

The effects of climate change do not cluster in ways that can be clearly linked by the nonscientist (and in some cases, even by scientists) to climate-change dynamics. The environmental and social impacts are and will be unevenly distributed, even within countries. So it is not uncommon that people living in different regions of a country differ in their views of what if anything is going on and what if anything needs to be done.

Preview of the Book

This book responds to these challenges. The authors bring the how, what, and why of climate change from the laboratory to the living room. *Climate Change* summarizes in understandable terms what science knows about climate change and addresses how that knowledge has been used and can be turned to action by government and business. The book also recommends ways to further the public's understanding of this complex international environmental challenge and to affect public opinion in ways that may drive policy and actions.

The book first offers a primer on global climate change. Chapter 2 explains the nuts and bolts of climate, the greenhouse effect, and historical discoveries of their interaction. Next, in chapter 3, we summarize the effects of climate change on the world, on regions, and on states. Here we describe how people, plants, animals, crops, and the natural environment are all affected by climate change. Adding a science historian's perspective, Naomi Oreskes explains in chapter 4 the nature and nurture of consensus in the climate-change debate and asks how we know that we're not wrong and whether the contrarians might yet carry the day.

Following these science-based chapters, in chapter 5 we explore world responses from the public and private sectors. How have international scientific and legal organizations reacted? What has been the U.S. position, and how has it changed? How have states and businesses large and small responded? Can New Jersey, California, and other states and BP, GE, and other major companies mitigate climate-change effects? The book then turns squarely to the question of how climate science is communicated to us, our children, and our grandchildren. In chapter 6, *New York Times* writer Andrew C. Revkin shares his experiences and ideas for improvement. An important factor in societal decisions about action and nonaction is the manner in which scientific information is understood. Richard Matthew addresses in chapter 7 the effect of climate change on "other people's children," especially in developing countries but also touching on areas at risk in the United States. Our concluding chapter takes readers to next steps in thinking about climate change and in acting on the science, the costs and benefits of actions of various sorts, the policy responses, and the roles that are appropriate to be taken by governments, businesses, and citizens.

Notes

1. According to the Public Policy Institute of California's *Special Survey on Californians and the Environment* (Baldassare 2003): "Two in three Californians (68 percent) believe that increased carbon dioxide and other gases released in the atmosphere will, if unchecked, lead to global warming. Forty-five percent of state residents—and 54 percent of those ages 18–34—believe that global warming will pose a serious threat to them in their lifetime. Nearly three of four (73 percent) believe that immediate steps should be taken to counter the effects of global climate change. What are they willing to do? Majorities say they are willing to make major lifestyle changes to address the problem (69 percent), believe that the federal government should set new legally binding industrial standards to limit emissions thought to cause global warming (66 percent), and that the federal government should work with other nations to set standards for the reduction of greenhouse gases (52 percent). Democrats (77 percent) are more likely

than Republicans (59 percent) to believe that global warming exists." National data on public opinion are summarized in chapter 5.

2. This friend attended the 2003 program sponsored by the Newkirk Center for Science and Society, which is the basis for several of the chapters in this book, and was impressed with what he learned.

References

Baldassare, Mark. 2003. *PPIC statewide survey: Special survey on Californians and the environment.* Public Policy Institute of California. ⟨http://www.ppic.org/content/pubs/survey/S_703MBS.pdf⟩ (accessed December 4, 2006).

McKibben, Bill. 2003. Worried? Us? *Granta* 83 (October 30). ⟨http://www.granta.com/extracts/2032⟩ (accessed December 4, 2006).

Revkin, Andrew C. 2004. Computers add sophistication, but don't resolve climate debate. *New York Times*, August 31, D-3.

Revkin, Andrew C. 2005. Bush aide edited climate reports. *New York Times*, June 8, A-1.

2

A Primer on Global Climate Change and Its Likely Impacts

John Abatzoglou, Joseph F. C. DiMento, Pamela Doughman, and Stefano Nespor

Greenhouse gases are accumulating in Earth's atmosphere as a result of human activities, causing surface air temperatures and subsurface ocean temperatures to rise.

—National Academy of Sciences (2001, 1)

Human activities over the past one hundred and fifty years have been changing the delicate chemical balance of the earth's atmosphere. Today, scientists are attempting to describe and quantify the effects of these activities on our climate, specifically as they relate to large changes in temperature and precipitation, the magnitude and frequency of extreme weather events, and changes in sea level—all of which will have countless unknown consequences on the planet.

In this chapter, we present a primer on climate change. Climate forms the basis for life to exist on earth. The development of life forms and human civilization would not have been possible without a complex climate system. To most people, climate seems a simple, intuitive process. But the climate system involves many physical, chemical, and biological interactions—among air, water, ice, land surface, plants, and animals. Although fully addressing these interactions would require several volumes, here we synthesize the important elements behind the science of climate and climate change. The complex nature of the earth's climate system presents a difficult scientific challenge. Researchers must identify the processes that drive the system, document past and current changes in climate, and project how natural variability and human activity may alter future climates.

Climate science is a rapidly evolving field where groundbreaking discoveries are made on a regular basis. Although scientists may never fully grasp every detail that comprises the climate system, we summarize here several key elements to give readers a background on climate science and climate change. First, we distinguish the commonly misused terms of *climate* and *weather*. Second, we introduce the important components and interactions that drive the climate system. Here we address the importance of the greenhouse effect and introduce the phenomenon of human-made changes in atmospheric greenhouse gases and *aerosols* (suspended solid and liquid particles). The chapter then compares the current climate situation with records from the past and uses this information to help predict what will happen to climate in the future.

Climate and Weather

Weather and climate are different concepts. *Weather* refers to atmospheric conditions for an individual event or time. For example, if someone asks you, "How's the weather?" you can simply go outside and answer this question easily. We are concerned with how meteorological conditions may affect us over the next week or so, and we refer to weather forecasts as a guide. Weather forecasting involves modeling the circulation of atmospheric conditions through time by employing a series of mathematical equations. Although the accuracy of weather forecasts has improved over the past couple decades because of advances in observation, analysis, and modeling, the current limit of weather forecast skill extends to about ten days.

Weather predictability is not reliable beyond this period, chiefly because (as MIT meteorologist Edward Lorenz first noted) even minute inaccuracies in initial observations (for example, in temperatures or winds) that are placed into a forecast model will grow exponentially in time. This is *chaos theory*, often referred to as the "butterfly effect," after the idea that a butterfly flapping its wings in Brazil could cascade to larger scales and set off a tornado in Texas. Unfortunately, we are currently unable to monitor the movement of each of the world's butterflies, and current forecast models tend to develop large errors.

Since this book addresses the issues of climate and anticipated climate change, a more appropriate question to ask might be, "How's the climate?" Mark Twain provides us with a good starting point: "Climate is what you expect; weather is what you get." *Climate* refers to an average of weather conditions over an extended period of time (months, years, or centuries). Meteorological elements that characterize climate include temperature, precipitation, cloud cover, humidity, and wind patterns. Climate ultimately influences human culture—from the articles of clothing we wear and the recreational activities we engage in to more pressing issues such as our food, water, and energy demands. So while we may not address climate on a daily basis, it truly dictates our way of life.

Although *weather* and *climate* are distinctly separate terms, just as there are daily fluctuations in atmospheric conditions (weather), there are also long-term fluctuations in climate. Recorded climate variations have caused or contributed to ecological adaptations, migrations, catastrophes, and successes.

About fifteen thousand years ago, the most recent ice age ended, and the glaciers covering North America and Europe began to recede. Nearly thirteen thousand years ago, during a period known as the Younger Dryas, glacial conditions rapidly returned to the high latitudes of North America, and average temperatures in northern Europe dropped nearly 5°C (9°F). Archaeologists note that the onset of this cooler and drier period coincided with the beginning of agriculture in northern Mesopotamia (Calvin 2002). Fluctuations in the climate likely had an adverse impact on the food supply of hunter-gatherers, creating an incentive for developing agriculture as a more stable and reliable food source. A couple of relatively minor climate variations during the last millennia, the Medieval Warm Period and the Little Ice Age, had dramatic impacts on the population of Europe. During the Medieval Warm Period (in the tenth to fourteenth centuries), a small increase in temperature was concentrated primarily over the North Atlantic basin and allowed the Vikings to colonize Greenland. However, the onset of the Little Ice Age (in the fifteenth to nineteenth centuries) ushered in cooler temperatures, which resulted in the collapse of these colonies. In addition, temperatures that were 2°C (3.6°F) cooler than today across much of Europe significantly reduced agricultural productivity and led to increased rates of starvation and an overall deterioration in human health. Some believe that these changes exacerbated the plague that ravaged Europe during this period.

On much shorter timescales, phenomena such as El Niño–Southern Oscillation and tropical volcanic eruptions have important but short-lived impacts on global climate. El Niño–Southern Oscillation is a atmosphere and ocean phenomena that channels interannual (year-to-year) fluctuations in ocean temperatures over the tropical east Pacific into global fluctuations in climate. For example, during winters when ocean surface temperatures off the coast of Peru are unusually warm, the Southern half of the United States generally sees cooler and wetter than normal winters while the Northern half of the country from the Great Lakes westward to the Pacific Northwest generally sees warmer than normal conditions (National Oceanic and Atmospheric Administration 1994). Volcanic eruptions have long been implicated with climate variations. Following an eruption, huge quantities of gases and aerosols are injected high into the atmosphere, where they form a cloud that effectively shields the earth from the sun's rays and leads to a short-lived (one to two years) global cooling. For example, it has been suggested that the eruption of Mt. Etna on the island of Sicily in 44 BCE cooled the planet and led to crop failures and eventual famine in Rome and Egypt (Robock 2002).

Earth's Climate System

The earth's climate system can be thought of as an elaborate balancing act of energy, water, and chemistry involving the atmosphere, oceans, ice masses, biosphere, and land surface. Although generated nearly 150 million kilometers away, radiation from the sun provides our planet with the energy that forms the basis of climate and makes life possible. The sun's energy propagates toward the earth as solar radiation. Solar radiation directed toward the earth carries energy amounting to 342 watts for every square meter of the surface of the earth. This is equivalent to six 60-watt light bulbs shining year-round for every square meter and represents nearly ten thousand times the energy consumed by humankind. However, not all the energy emitted from the sun actually reaches the earth's surface. Roughly 30 percent of the incoming solar radiation is reflected back to space off of bright surfaces on the planet, including snow cover and sand, and bright surfaces in the atmosphere, including clouds. This reflection to space happens through the same phenomenon that keeps a white car, which readily reflects solar radiation, much cooler on a hot day than a black car, which readily absorbs solar radiation. The remaining two thirds of solar radiation that penetrates the atmosphere and reaches the earth's surface heats the land and oceans.

Just as the sun emits energy, so does the earth. Emitted energy, or *radiation*, from both the sun and the earth travels in the form of waves that are similar to the waves moving across the surface of a pond. However, the energies emitted by the sun and earth are different due to the large differences in temperature between the two bodies. They emit radiation at distinctly different *wavelengths* (the distance between adjacent crests in the wave). While the hot sun emits energy at short wavelengths (referred to as *shortwave radiation*), the much cooler earth emits radiation at longer wavelengths (referred to as *longwave, or thermal, radiation*).

If the earth is to maintain a constant temperature, the amount of radiation received by the earth from the sun must be balanced by an equal amount of radiation emitted from the earth back to space. Without an atmosphere, the only means by which the earth's temperature could change would be through changes in solar input. According to simple energy-balance calculations, the average temperature of the earth should be −18°C (0°F). Fortunately, the earth has an atmosphere that acts like a blanket that traps much of the outgoing thermal radiation emitted by the earth's surface but allows most of the solar radiation to pass through. Certain trace gases in the earth's atmosphere, called *greenhouse gases*, selectively

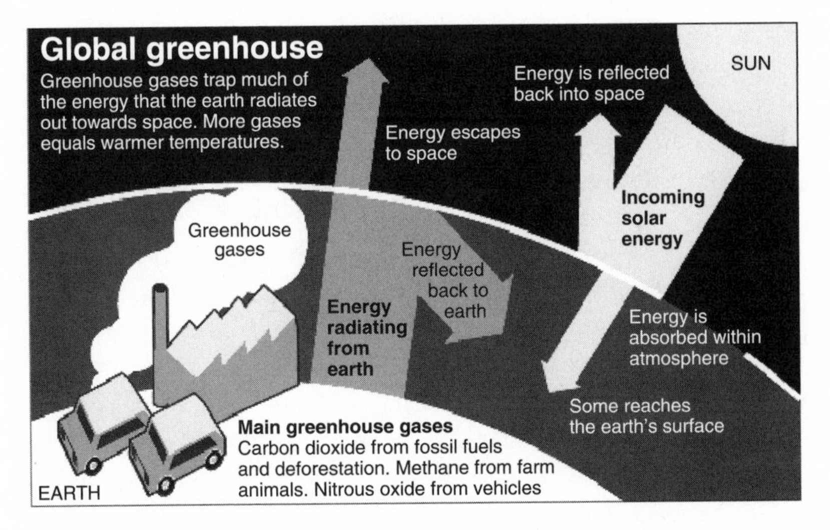

Figure 2.1
Fundamental dynamics of the greenhouse effect

absorb and trap these longer wavelengths of energy emitted by the earth and then reemit them back to the earth's surface. This allows for a significant warming of the earth's surface and its lower atmosphere.

Water vapor and clouds account for much of the natural greenhouse effect. The warming effect of clouds can be observed during winter nights. A cloudy winter night is often much warmer than a clear winter night, since clouds and water vapor trap the heat radiating from the surface and keep surface temperatures from dropping as much as they would on a clear night. Overall, the greenhouse effect allows the average surface temperature of the earth to warm from a frigid −18°C (0°F) to a more comfortable 15°C (59°F). Thus, the chemical makeup of the atmosphere is crucial in establishing a climate that is hospitable to life (figure 2.1).

This description of energy balance is relevant for the earth as a whole, but significant variations in energy input and output exist across the globe. Although all locales on the face of the earth receive the same duration of sunlight over the course of a year, solar radiation is much more intense near the equator than near the poles. Consider an example where you shine a flashlight, representing the sun, at a sheet of paper, representing the earth's surface. The intensity of light on the paper changes dramatically as you tilt the paper toward and away from the beam of the flashlight. The intensity of light on the paper is strongest when the beam is directed perpendicular to the paper, which is analogous to the sun when it is directly overhead. Due to the curvature of the earth's surface and the tilt of the earth's axis, the sun's rays are directly overhead in tropical locales throughout the year. In contrast, the sun is much lower on the horizon over polar regions, analogous to your flashlight beam striking the paper at a much lower angle. However, the rate at which earth emits energy (radiation) to space, which depends on temperature, does not differ dramatically from the equator to the poles. As a result, there is a net loss of radiation near the poles because there is less solar radiation coming in than thermal radiation being emitted. Near the equator, the opposite occurs, and there is a net gain of radiation. To maintain a steady climate, the ocean and atmosphere must transport excess heat from the tropics to the heat-deficient polar regions. That heat is moved by the action of winds and ocean currents. Without a dynamic atmosphere and ocean system, there would be a precipitous cooling of the poles and a dramatic warming of the tropics. The circulation of our atmosphere and oceans redistributes this energy imbalance in setting the earth's climate, thus making life more habitable on much of the globe.

The atmosphere responds to unequal heating of the earth's surface by generating atmospheric motion. *Atmospheric circulation* redistributes heat around the globe in an attempt to create energy balance. Atmospheric circulation responds quickly to radiation. A small-scale example of such a response is the sea breeze. On a hot day, the land surface heats up much faster than does the ocean surface. The warmer air over land becomes less dense and begins to rise. As this air mass rises, the cooler air over the ocean flows inland to replace the rising warm air. As a result, the sea breeze cools inland locations in an attempt to offset temperature differences. On larger scales, the general circulation of the atmosphere arises in a similar, more involved process to counter global-scale imbalances in heating. Consider, for example, the intense surface heating of the tropics as an analog of the intense heating over the land in the above example. Warm air in the tropics is forced to rise upward and eventually poleward. This redistributes energy around the globe, bringing warm air to higher latitudes and cool air to lower latitudes and ultimately defining climate as we know it.

Oceans are a key component of the climate system. Oceans contain 97 percent of the planet's water and cover over 70 percent of the earth's surface. Among the unique properties of water is its ability to store vast quantities of heat. As surface water in the tropics is heated, large-scale ocean currents, driven by atmospheric circulation patterns, transport heat poleward. This process parallels the atmosphere's redistribution of energy across the globe, but it operates on much longer timescales. Consider, for example, the circulation of the North Atlantic basin. The northward-flowing Gulf Stream transports warm water from the Gulf of Mexico toward northern Europe. It is believed that the Gulf Stream is partially responsible for the relatively mild climates of northwestern Europe, although the large-scale atmospheric circulation likely plays a dominant role (Seager 2006). The ocean also plays a role in determining the chemical composition of the atmosphere because it absorbs and releases gases. The most recognized example is the evaporation and precipitation of water vapor through what is called the *hydrologic cycle*. However, the ocean also emits and absorbs large quantities of atmospheric carbon dioxide (CO_2).

The *cryosphere* comprises all frozen water, including the Greenland and Antarctic ice sheets, sea ice in the Arctic and Southern Oceans, and all other snow- and ice-covered surfaces. The cryosphere represents only about 2 percent of the water on our planet, but it is important to the climate system because it reflects incoming solar radiation. Large ice sheets reflect between 80 and 90 percent of solar radiation, allowing very little radiant energy to warm the surface. The growth of the ice sheet during a glacial period acts as a positive feedback on the climate system. A *positive feedback* involves a change in one element or variable (in this case, the growth of the ice sheet) that provokes a change in a system (in this case, enhanced reflection resulting in cooling). This cooling then "feeds back" to amplify the effect of the initial change. In this case, the cooling would lead to a further expansion of the ice sheet (and so on). Conversely, as the ice sheet recedes during a warming phase, a greater amount of energy reaches the earth's surface, accelerating the warming and the melting of the ice sheet. Throughout this book the term *forcing* or *climate forcing* refers to any imposed mechanism that forces climate to change.

The land surface and the biosphere also contribute to the climate system. The large-scale configuration of the continents—including altitude, proximity to the ocean, and prominent mountain ranges—alters atmospheric and oceanic circulation patterns. In addition, variations in land surface alter the exchange of both heat and water, thus affecting local and regional climates. These factors combine to produce great differences in regional climates. San Francisco and Washington, DC, for example, are located at approximately the same latitude but experience strikingly different climates.

The land surface and the biosphere both affect and are affected by atmospheric temperature and humidity and can alter the amount of solar radiation reflected back to space. Vegetation also plays a key role in the *carbon cycle*, which is the exchange of carbon among atmosphere, ocean, and land (biosphere included). Plants are active participants in the carbon

Box 2.1
Definition of Climate

The word *climate* is derived from the Greek word *klima*, a term that refers to the inclination of the sun's rays to the earth's surface. The earth's various climates reflect the tilt of the planet's land surface in its orbit around the sun. The Greek geographer Ptolemy proposed that latitudinal changes in tilt (and therefore climate) affected the length of the day and the brightness of the sun—ultimately altering livability on earth. In addition, changes in regional distribution of solar heating drive climate and result in regional variations in temperature, wind, and precipitation patterns. The regional climates that are commonly accepted today come from Wildimir Koppen's subdivision of the earth based on temperature, precipitation, and distribution of natural vegetation. Climate ultimately represents the complex web of factors that define the atmospheric conditions in a determined geographical area for a prolonged period of time (years, decades, centuries, and geological eras). Climate takes into account the average weather over an extended period; it can be seen as an *integrator* of weather. In contrast, weather accounts for the variability in atmospheric conditions that occur from day to day.

cycle as they absorb CO_2 through photosynthesis and expel CO_2 through respiration. It is currently thought that plants are net absorbers of atmospheric CO_2 (Sarmiento and Gruber 2002). However, changes in land use—such as deforestation and subsequent burning or decomposition of forest material—release an abundance of stored carbon into the atmosphere and obstruct processes that remove greenhouse gases and aerosols (scientists call these removal processes *sinks*).

The Importance of Greenhouse Gases to Climate

In 1824, the French scientist Joseph Fourier hypothesized that the average temperature of the planet is warmer because of the existence of the earth's atmosphere. He claimed that the warming effect of the atmosphere on the earth's surface was similar to how a plant warms when it is encased in a house of glass. Fourier called this phenomenon the *greenhouse effect*.

The composition of the earth's atmosphere governs the climate of the planet and establishes conditions vital for life. Although the atmosphere is primarily composed of nitrogen (78 percent) and oxygen (21 percent), these gases do not interact with the longwave thermal radiation emitted by the earth. This task is left to the greenhouse gases, which account for less than 3 percent of the atmosphere. Greenhouse gases—including carbon dioxide (CO_2), methane (CH_4), nitrous oxide (N_2O), halocarbons, ozone (O_3), and water vapor (H_2O)—are very effective at absorbing thermal radiation expelled from the earth's surface.

After greenhouse gases absorb thermal radiation emitted from the earth's surface, they reradiate this energy back to the surface of the earth, which warms the earth in the same way that a blanket traps body heat on a cold night. While greenhouse gases absorb and emit thermal radiation, they are essentially transparent to solar radiation and allow additional heat into earth's atmosphere, where it is trapped by the greenhouse gases. This system permits hospitable conditions for life on the earth's surface. However, small changes in the concentrations of these gases can drastically alter the heat-trapping capabilities of our atmosphere, resulting in acute changes in climate with serious consequences for life on earth.

The four planets that are closest to the sun are Mercury, Venus, Earth, and Mars. Earth's two neighboring planets, Venus and Mars, offer good examples of how changes in atmospheric composition can lead to changes in surface temperatures. Although Venus is closer than the earth is to the sun and thus receives a greater amount of incoming solar radiation, thick clouds engulf the planet and reflect nearly 75 percent of this radiation (compared to 30 percent for earth). The atmosphere of Venus is rich in greenhouse gases, with carbon dioxide accounting for 97 percent of it. As a result, the thick Venusian atmosphere is highly effective at trapping thermal radiation from escaping to space. The large amounts of greenhouse gases in the atmosphere reradiate the trapped heat back to the surface of the planet, resulting in average surface temperatures of 470°C (878°F). In contrast, Mars has a very thin atmosphere with a minimal greenhouse effect. As a result, most of the heat radiated from the surface of Mars escapes to space, and the average surface temperature on Mars is about −60°C (−76°F).

The ability of greenhouse gases to warm the surface of the planet depends on three main factors: their efficiency in absorbing heat, their total atmospheric quantities, and their atmospheric lifetimes (or the amount of time they remain in the atmosphere).

Efficiency of Greenhouse Gases in Absorbing Heat

Greenhouse gases are defined by their ability to absorb thermal radiation emitted by the earth. Different molecular structures of the gases lead to differences in their ability to absorb radiation. Scientists estimate the heat-trapping efficiency of the different greenhouse gases using an index called the *global warming potential* (GWP). This represents the ratio of energy reemitted to the earth's surface during a year for a given gas compared to that of the same mass of CO_2. The global warming potential of CO_2 is defined as 1. By comparison, methane has a GWP of 21, meaning that a given mass of methane can heat the planet twenty-one times as much as the same mass of CO_2. Other greenhouses gases have even larger GWPs. Nitrous oxide and halocarbons have GWPs of 300 and over 5,000, respectively. So although carbon dioxide is notorious for its role in global warming, other less well-known greenhouse gases also play potent roles in the process.

Quantities of Greenhouse Gases

Carbon dioxide is a naturally occurring greenhouse gas that cycles through reservoirs in the land, ocean, atmosphere, and vegetation. Atmospheric CO_2 has been maintained between 180 and 280 ppm (lower during glacial periods and higher during interglacial periods) over the last 650,000 years. *Anthropogenic*, or human-made, CO_2 began to be emitted to the atmosphere when people started to burn wood and fossil fuels. Methane is produced naturally, but its atmospheric concentrations have been augmented by agricultural processes (such as rice cultivation, use of fertilizers, and cattle farming) and industrial activities. Similarly, the amount of nitrous oxide in the atmosphere has increased as a result of agricultural soil

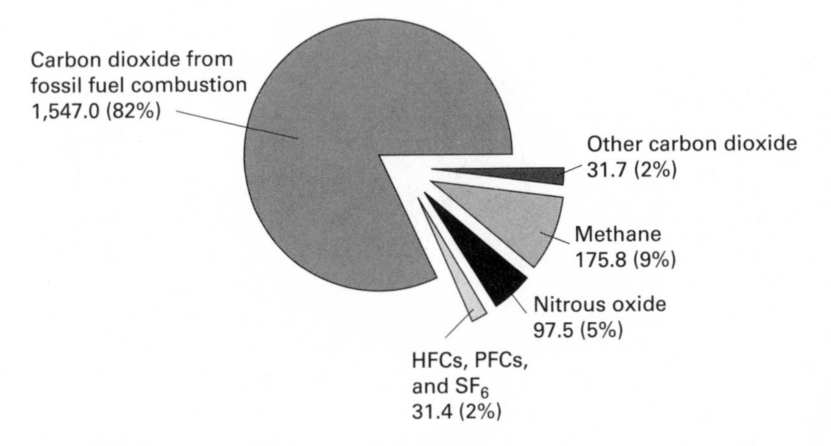

Figure 2.2
United States emissions of greenhouse gases by type
Source: Energy Information Administration (2002).

management, fossil-fuel burning, and biomass burning. The proportions of emissions of these gases for the United States are displayed in figure 2.2, and global emissions in figure 2.3.

Halocarbons are potent greenhouse gases that do not exist in nature. They are gases that are manufactured for use in refrigeration units and foaming agents. Alternative halocarbon compounds, such as hydrofluorocarbons and perfluorocarbons (PFCs), were introduced as substitutes for chlorofluorocarbons (CFCs), which are potent greenhouse gases that also destroy the ozone layer. CFCs act in a series of chemical reactions to destroy the natural shield of ozone high in the atmosphere, which protects life from the infiltration of dangerous ultraviolet rays. Those rays can harm human health (by leading to cataracts, suppressed immune systems, and some forms of skin cancer), plant species, and oceanic plankton (World Meteorological Organization 2002). Because of the overwhelming

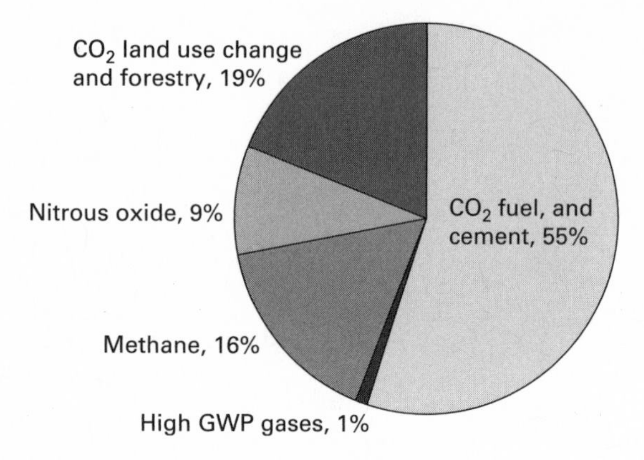

Figure 2.3
Global greenhouse gas emissions, 2000

worldwide effort to curtail the usage of CFCs since the Montreal Protocol of 1987, the earth's ozone layer is slowly recovering. The success with CFCs provides hope that we are able to alter behavioral patterns to repair the damage we cause to earth's atmosphere.

Lifetimes of Greenhouse Gases

Most greenhouse gases have lifetimes of decades to centuries. This means that a gaseous molecule may remain in the atmosphere for as long as two hundred years, mixing throughout the atmosphere. Water vapor—with a short lifetime of a few days to weeks—is not well mixed in the atmosphere, and many locations have high humidity, cloud cover, and rainfall, while other locales are dry and cloud free. Most other greenhouse gases have long lifetimes and continually accumulate in the atmosphere, leading to important long-term implications for future climate changes.

Industry and Greenhouse Gases

Until a few centuries ago, the earth's radiative equilibrium and climate were maintained by the natural greenhouse effect. Atmospheric levels of carbon dioxide were balanced by the carbon cycle, and there was equality between sources and sinks. Carbon cycled through photosynthesis and respiration by land and sea flora, through air-sea exchanges (*fluxes*), and through "slow-turnover" geologic processes. Over millions of years, oceanic carbon is buried, and deep-sea sediments are recycled into the earth. Eventually, this carbon is reintroduced to the atmosphere—either violently by volcanic eruptions or mildly by the breakdown of exposed rocks. (Earth's largest reservoir for carbon is rock.) While important for the evolution of the atmosphere, the relatively slow pace of the geologic carbon cycle (millions of years) is important only in controlling long-term variations in levels of atmospheric carbon dioxide.

Atmospheric concentrations of other natural greenhouse gases, such as nitrous oxide and methane, were also maintained by a balanced cycling of nitrogen and carbon within the earth's climate system.

Scientists can quantify the composition of the atmosphere prior to the historical record by examining ice cores. Bubbles of air embedded within the ice cores that have been extracted from the Greenland and Antarctic ice sheets reveal a substantial amount of information on changes in climate. The cores tell us that from about 420,000 years ago until the beginning of the industrial revolution in the late 1700s, CO_2 varied from about 180 parts per million (ppm) to about 280 ppm (Siegenthaler et al. 2005). As of March 2006, there were about 381 ppm of CO_2 in the atmosphere, with levels of CO_2 rising

Box 2.2
Charles David Keeling, 1928–2005

In 1958, Charles David Keeling, a professor at the Scripps Institute of Oceanography at the University of California at San Diego, began to collect a continuous record of atmospheric carbon dioxide concentrations from towers on the Mauna Loa Observatory in Hawaii at an elevation of about 4,100 meters (13,450 feet). Prior to his observations, it was unclear whether increased CO_2 emissions would actually accumulate in the atmosphere or be absorbed by vegetation and the oceans. Keeling's nearly fifty years of measurements have confirmed an alarming increase in ambient CO_2 levels and are a leading piece of evidence of anthropogenic effects on the atmosphere. Keeling found a rise in atmospheric CO_2 from 316 ppm in 1958 to 376 ppm in 2003. The level in March 2006 was 381 ppm. In 2002, Keeling received the National Medal of Science from the National Science Foundation at the White House in recognition of his lifetime achievements. Many consider his work to be the single most important environmental dataset of the twentieth century.

about 1.6 ppm per year since 1980 (National Oceanic and Atmospheric Administration 2006). Measurements confirm a high correlation between fluctuations in atmospheric CO_2 levels and global temperature. These observations are consistent with the radiative impact on surface temperatures from changes in the greenhouse effect. The question is: How will these high levels of CO_2 affect our climate?

The industrial revolution marked a turning point in the balance of energy in the earth's climate system. The rise of industry and technology throughout the 1800s saw a massive increase in the burning of wood and coal, which released the greenhouse gases carbon dioxide, nitrous oxide, methane, and other halocarbons into the atmosphere. This industrial activity

Box 2.3
The Colonials

Colonials were sensitive to British criticisms of the cold American climate, and according to science historian James Fleming (1998), they argued that their climate was improving as the forests were cleared. In 1721, Cotton Mather believed that the North American region was getting warmer: "Our cold is much moderated since the opening and clearing of our woods, and the winds do not blow roughly as in the days of our fathers, when water, cast up into the air, would commonly be turned into ice before it came to the ground." Fleming (1998, 24) notes that Benjamin Franklin agreed, writing to Ezra Stiles in 1763 that "cleared land absorbs more heat and melts snow quicker," but concluded that many years of observations would be necessary to settle the issue of climatic change.

thus altered the carbon cycle and the natural equilibrium of the earth's climate system. At the same time, deforestation further exacerbated the balance by removing a potential sink for carbon. Today, human activity is responsible for releasing approximately 7 billion metric tons of carbon per year into the atmosphere, and the oceans and land biosphere absorb approximately 3 billion metric tons of that carbon (Shein 2005). Since CO_2 has a lifetime of over one hundred years, these emissions have been collecting for many years in the atmosphere. The atmospheric concentration of CO_2 has increased 36 percent from preindustrial times and is expected to continue rising in the foreseeable future.

Electricity generated by the burning of fossil fuels accounts for most of the more than 5.5 billion tons of human-made carbon dioxide released each year by the United States (Environmental Protection Agency 2006). In contrast, other energy sources—including nuclear, solar, wind, hydroelectric,

Box 2.4
Guy Stewart Callendar, 1898–1964

Beginning in 1938, the British engineer and scientist Guy Stewart Callendar identified important links between the burning of fossil fuels and global warming (Fleming 2007). He compiled weather data from stations around the world that indicated a climate warming trend of 0.5°C (0.9°F) in the early decades of the twentieth century. He investigated the carbon cycle, including natural and anthropogenic sources and sinks, and the role of glaciers in the earth's heat budget. His estimate of 290 ppm for the nineteenth-century background concentration of carbon dioxide is still valid, and he documented an increase of 10 percent between 1900 and 1935, which closely matched the amount of fuel burned. Based on new scientific understandings and calculations, Callendar established the CO_2 theory of climate change in its recognizably modern form, reviving it from its earlier unrealistic status. Today, the theory that global climate change can be attributed to an enhanced greenhouse effect linked to elevated levels of CO_2 in the atmosphere from anthropogenic sources, primarily from the combustion of fossil fuels, is called the *Callendar effect*.

biomass, and geothermal energy sources—emit minimal, if any, greenhouse gases.

The transportation sector is the second biggest source of carbon dioxide emissions in the United States. Every gallon of gasoline consumed releases about 9 kilograms (20 pounds) of CO_2 into the atmosphere. Fuel economies of many automobiles have improved dramatically over the past few decades because of technological improvements, but the carbon dioxide emitted from vehicles in the United States exceeds the total carbon dioxide emitted from India (even though the population of India is over three and a half times that of the United States).

In addition to CO_2 emissions in the electricity and transportation sectors, emissions of less notorious yet equally dangerous greenhouse gases are linked to climate change. About two-thirds of present methane emissions are attributable to human activities, such as burning biomass, cultivating rice, creating landfills, and managing livestock. These methane releases from agricultural and natural sources are expected to be accelerated by changes in global mean temperature and moisture. For example, recent studies indicate that climate warming may melt high-latitude permafrost and thus accelerate releases of methane from peat bogs (Zimov, Schuur, and Chapin 2006). The switch from CFCs to other halocarbons has limited the destruction of the ozone layer, but these halocarbons remain active in enhancing the greenhouse effect and potentially harmful climate change.

Aerosols

In addition to gases such as carbon dioxide, methane, and water vapor, the atmosphere contains suspended solid and liquid particles, called *aerosols*. Aerosols can range in size from tiny molecular clusters to particles visible to the human eye. The principal sources of aerosols are fossil-fuel combustion, biomass burning, desert dust, volcanoes, and sea spray. The most straightforward climate effect of aerosols is their ability to alter the amount of incoming solar radiation that reaches the earth's surface. Much like clouds, light-colored aerosols reflect incoming solar radiation, thereby decreasing the amount of energy that reaches the earth's surface. These aerosols cool the planet, and some researchers have promoted them as the antidote to greenhouse gases.

Over the last two hundred years, sulfate aerosols, a by-product of emissions from fossil-fuel combustion, have increased enormously. In parts of the world that consume large amounts of high-sulfur coal, such as China and central Europe, sulfate emissions could in the short term actually offset the warming effects of high levels of greenhouse gases. But emitting these aerosols to attenuate the enhanced greenhouse effect is inadvisable for several reasons. First, sulfates combine with water vapor to form sulfuric acid, the principle component in environmentally harmful acid rain. Second, unlike the long-lived greenhouse gases, aerosols have a relatively short residence time of days to weeks, so massive and continual emissions of sulfates would be required to counter the effects of localized warming. Such short atmospheric lifetimes mean that aerosol concentrations are highly localized: they affect only areas near their emission sources, while long-lived greenhouse gases are diffused across the globe. Finally, loading the atmosphere with aerosols has an unknown effect. Because aerosols interact with cloud and precipitation formation, conducting a global experiment of this type to combat global warming would be unsound.

Views on human contributions to the greenhouse effect have been a major source of political controversy in the United States and other countries. The fundamental issue is whether climate changes might result from natural variations and not significantly from human activities. Numerous studies have been conducted to determine the extent to which climate changes realized during the twentieth century fall within the natural realm of variability (Karl and Trenberth 2003). Models that account solely for the full range of natural forcing (that is, volcanoes, internal climate variability, and solar variability) are unable to replicate the observed twentieth-century warming trend. However, after adding anthropogenic forcing (such as greenhouse gases and aerosols), models show a warming trend that mirrors the trend found in observations, which does not appear consistent with the alternative explanations cited by several climate contrarians. Model results provide solid evidence that a majority of the warming observed during the twentieth century is tied directly to increases in levels of atmospheric greenhouse gases.

This evidence was presented to the White House in spring 2001 in a report that concluded that "Greenhouse gases are accumulating in Earth's atmosphere as a result of human activities, causing surface air temperatures and subsurface ocean temperatures to rise. . . . The changes observed over the last several decades are likely mostly due to human activities" (National Academy of Sciences 2001, 1). The report went on to say that over the past million years transitions between glacial and interglacial periods have exceeded 10°C (18°F) regionally. Determining global (rather than local and regional) temperature changes is more challenging, but "global warming rates as large 0.2°C (0.36°F) per century may have occurred during retreat of the glaciers following the most recent ice age" (National Academy of Sciences 2001). A warming of 2°C (3.6°F) per one thousand years is a third of the rate of the observed 0.8°C (1.4°F) warming over the last one hundred years. In fact, the current rate of change exceeds the largest warming rate seen in earth's climate history.

Climate Change as an Environmental Problem

Climate change was recognized as a global environmental problem for the first time in the late 1970s. James Hansen, an American climatologist, advanced the hypothesis that the

Box 2.5
James Hansen

Although James Hansen's (b. 1941) early research focused on the properties of the atmosphere of Venus, he soon was drawn to the exciting and innovative research being done with climate change on earth—specifically, the study of human effects on climate. Since then, he has deepened knowledge of many aspects of the climate-change issue that have caught the attention of world public opinion. He has developed a number of climate models that help to explain the present intensity of climate change and has formulated scenarios that take into account the impact of human activities on climate. Based on his research, Hansen concludes that "although there are confusing contributions to the topic, there is no doubt that the Earth has warmed during the last century. Moreover, by now there is decisive evidence that carbon dioxide and other gases such as methane and atmospheric pollution have a direct causal connection with this warming." Hansen is currently the director of the NASA Goddard Institute for Space Studies in New York.

burning of fossil fuels over a long period of time was slowly but progressively heating the planet. At the first World Conference on Climate, held in Geneva in 1979, many scientists warned that human activity could produce changes in climate that would harm people and the environment. The conference attendees invited all heads of state to heed their warnings about climate change and enact "necessary policies for the well-being of humanity." In chapter 4, we describe the worldwide response to this invitation and later warnings.

Although Hansen helped raise awareness of the problem of climate change, his thesis was not new. In the late 1800s and early 1900s, the future Swedish Nobel laureate Svante Arrhenius (1896, 1908) calculated an atmospheric warming provoked by an increase in emissions of carbon dioxide from the

Box 2.6
Svante Arrhenius, 1859–1927

"Is the mean temperature of the ground in any way influenced by the presence of the heat-absorbing gases in the atmosphere?" In 1895, the Swedish chemist Svante Arrhenius presented an answer to the Stockholm Physical Society (Fleming 1998). In a work entitled "On the Influence of Carbonic Acid in the Air upon the Temperature of the Ground," Arrhenius formulated a heat budget for the planet in which changes in the atmospheric levels of carbon dioxide are matched by changes in surface temperature. With this model, he concluded that the "temperature of the Arctic region would increase by 8° or 9°C (14°–16°F) if there were an increase of CO_2 in the atmosphere of 2.5 to 3 times over the then present values." Conversely, he argued that to achieve ice-age temperatures in the temperate zones, levels of atmospheric CO_2 would have to be reduced by about 40 percent. As Elisabeth Crawford has shown, Arrhenius did not write his Physical Society essay because of any great concern for increasing levels of CO_2 caused by the burning of fossil fuels. Instead, he was attempting to explain temperature changes at high latitudes that could account for the onset of ice ages and interglacial periods. His essay represented one of a number of contributions to the ongoing quest at the Stockholm Physical Society to develop a cosmic physics linking the heavens and the earth (Crawford 1996).

Arrhenius immersed himself in climate study. His popular book *Worlds in the Making* (1908) described the "hot-house theory" (greenhouse effect), calling on earlier studies and confirming that the earth's surface temperature would be about 30°C (55°F) cooler than it presently is without the effect of atmospheric gases. In 1904, Arrhenius noted that the increased production of CO_2 "by the advances of industry" could alter climate "to a noticeable degree in the course of a few centuries." However, he maintained that this increased use of fossil fuels could be advantageous for the climate, resulting in "ages with equable and better climates, especially as regards the colder regions of the Earth."

In 1899, Nils Ekholm, an associate of Arrhenius, pointed out that then present rates of the burning of pit coal could double

Box 2.6
(continued)

the concentration of atmospheric CO_2 and could "undoubtedly cause a very obvious rise of the mean temperature of the Earth." In *Worlds in the Making*, Arrhenius popularized the 1899 observation by Nils Ekholm. Arrhenius considered it likely that in future geological ages the earth would be "visited by a new ice period that will drive us from our temperate countries into the hotter climates of Africa." On the timescale of hundreds to thousands of years, however, Arrhenius speculated on a "virtuous circle" in which the burning of fossil fuels could help prevent a rapid return to the conditions of an ice age and could perhaps inaugurate a new carbon-linked age of enormous plant growth. Arrhenius's CO_2 theory of climate change fell out of scientific favor and was not revived in its modern form until the mid-1950s (Fleming 1998).

burning of coal and oil. His results were not substantially different from those obtained by present models. Beginning in the early 1980s, the international community, primarily the United Nations and governments of the industrialized nations, invested enormous sums of money in scientific research to explore this hypothesis.

Reconstructing Past Climate and Forecasting Future Climate

Numerous natural and anthropogenic factors have produced today's climate. The study of past climates (in a field called *paleoclimatology*) and present-day climate helps to describe the evolution of the earth's climate system and the forces that drive the earth's natural climate variability. If we can thoroughly understand past variations in the climate records, then we might be able to foresee the evolution of future climate.

Natural variations in climate—including oscillations in large-scale wind patterns, changes in oceanic circulation, and periodic fluctuations in the earth's orbit—occur on timescales ranging from years to decades to millennia. Several of these patterns are well accepted by the scientific community, but countless other modes of climate variability are not fully understood. The world's longest observational record of climate—that of central England—was begun in 1659 (Met Office 1997), and most reliable measurements of climate (and related phenomena) go back only to the nineteenth century.

Observations of climate change provide researchers with a mere sliver of earth's climate record, but it is possible to reconstruct climate variability beyond what is observed by studying substitutes—what are called *proxy indicators*. Proxy data are essentially "natural" recording systems of past climate. They are found in sediments, ice cores, tree rings, and corals. The chemical composition of the air, for example, can be documented in the bubbles of air imprisoned in glaciers on high mountain peaks and at the polar ice caps. Ice cores extracted from hundreds of meters below the surface reveal the chemical composition of the air as well as information on temperature, precipitation, and wind patterns from hundreds of thousands of years ago. By examining the rings of old trees, paleoclimatologists can infer temperature and precipitation information across much of the world. In addition, scientists can use fossilized pollen samples and soil sediments to identify vegetation that flourished millions of years ago.

These techniques indicate that high levels of CO_2 correspond to warm periods. One of the largest warming events in the geologic record occurred during the Eocene epoch some 50 million years ago, when global temperatures are estimated to have been nearly 7°C (12°F) warmer than present and levels of

CO_2 are thought to have been three times their present concentrations. At those temperatures, palm trees could thrive in Wyoming, and crocodiles could inhabit the Arctic (Greenwood and Wing 1995).

Abrupt Change?

Climatologists generally place the climate record within a context of continuous change. Most simulations of future climate changes focus on a smooth transition from what scientists call one *equilibrium climate* to another equilibrium climate—for example, from 280 parts per million CO_2 to two times that level. But another future is possible: one of "rapid, large, and unexpected impacts on local, regional and global scales" (McCarthy et al. 2001, 946). These effects would certainly not be of the kind depicted in the 2004 movie *The Day after Tomorrow*, but they would take place over decades, which is rapid relative to geological time. The climate record in polar ice cores provides evidence that major climate change can occur in decades or even years (Severinghaus et al. 1998), and earth would suffer much fiercer consequences if a 2°C (4°F) change occurs over a twenty-year period rather than over a two-hundred-year period.

According to general models of atmosphere and ocean circulation, such an abrupt change in climate could result in the shutting down of oceanic circulation (called *thermohaline*). The three-dimensional ocean movement that encompasses the planet is driven by slight differences in *water density*, which is determined by both temperature and *salinity*, or salt content. At the surface of the ocean, winds also guide oceanic currents. Oceanic circulation (often called the *oceanic conveyor belt*) helps redistribute heat across the globe. As part of this oceanic conveyor belt, the Gulf Stream pulls warm salty water away from the low latitudes and toward the high latitudes, releasing a tremendous amount of heat into the North Atlantic and contributing to the relatively mild conditions across the northern latitudes of Europe (McCarthy et al. 2001, 948). A sudden collapse of oceanic circulation and the Gulf Stream would drasticly change heat transport and likely redefine climate for much of the globe. Although most models do not depict the thermohaline circulation shutting down before 2100, some show a possible collapse after 2100, especially if climate change occurs abruptly (McCarthy et al. 2001, 15). Oceanic circulation is more apt to be maintained if any warming or cooling is of a gradual nature. In turn, if the thermohaline circulation is maintained, it will inhibit abrupt climate change (McCarthy et al. 2001, 17).

Paleoclimatologists tell us that ocean-circulation patterns may have been disrupted about 8,200 years ago. One notion is that a melting ice sheet released a gigantic amount of fresh water over portions of North America. The influx of fresh water into the North Atlantic basin diluted the salinity and density of ocean waters, which subsequently inhibited ocean-atmosphere circulation in the North Atlantic. This reduced atmospheric temperatures by 4°C (7°F), producing a cold snap that lasted up to two hundred years. Climate models are able to replicate a slowing of the Gulf Stream and the consequent decrease in temperatures across the Northern Hemisphere (Davidson 2004).

Scientists who study changes in the earth system cannot follow classical experimental methods because we only have one earth: we do not have a second one without elevated levels of greenhouse gases in its atmosphere to observe for comparison. Instead, our control standard for checking our hypotheses is a computer model that captures much of the complexity of the earth system. Over the last two decades, improvements in observational and modeling techniques and in informational and computer sciences have led to a fuller understanding of the earth's climate system. Without the immense computing power now available and the vast infrastructure of widely available datasets, many advances in understanding could not have been realized. Today's attempts to integrate all available data into computer models are helping scientists to reconstruct variations in global climate over the last millennium and to predict future climate.

Uncertainties

Climate models have improved greatly in recent decades, but the best models still cannot accurately project the future climate of our planet. Limitations in the current understanding of natural processes, limited computing power, and the ever-confounding chaos theory all add degrees of uncertainty. To predict climate change, scientists must consider a plausible range of social, economic, and scientific conditions that might impact climate over the next hundred years. They need to account for changes in population and for a wide span of mitigation possibilities before they can estimate greenhouse-gas emissions. Scenarios range from the optimistic (where population levels stabilize and levels of greenhouse-gas emissions are reduced) to the pessimistic (where population and greenhouse-gas emissions continue to grow). These scenarios are then run through several different climate models, and the results provide scientists with both a ballpark figure and a measure of certainty. *Climate Change 2007* (Alley et al. 2007), a report that was written by hundreds of scientists for the Intergovernmental Panel on Climate Change, summarizes the current understanding of climate change and predicts the ways that climate may evolve in the near future.

Scientists have used many sources of data to investigate the dynamics of the climate system. These datasets describe the upper levels of the atmosphere, the depths of the oceans, the subtropical deserts, and the frozen tundra. They have been acquired via satellites, radar, weather balloons, submarines, and coring drills. Nonetheless, climate remains an unsolved puzzle. Climate forecasters suggest that clouds constitute the largest uncertainty for predictions about climate change. Clouds alter the earth's net energy balance since they both absorb outgoing thermal radiation (and thus help to warm the planet) and reflect incoming solar radiation (thus helping to cool the planet). Future changes in cloud distribution, quantity, and type will have important consequences for future climate change. As the National Academy of Science has concluded, "climate sensitivity" and feedbacks are needed to help us understand climate change. The Academy's 2001 report explains that water, whether dissolved in the atmosphere or frozen into great expanses of reflective surfaces, can greatly increase warming from greenhouse gases and that under certain conditions water vapor can increase warming by a factor of 1.6. Combined with the differences in the amount of energy reflected by ice instead of by water or land, the feedback effects from water can amplify global warming two to four times.

Conclusion

We have reviewed the basics of climate science and climate change, emphasizing how the earth's climate system balances both energy and the composition of the atmosphere to sustain life on the planet. This natural balance in the climate system is being threatened by an abrupt rise in greenhouse gases since the industrial revolution. Ambient levels of carbon dioxide increased from about 280 ppm in the mid-1800s to nearly

380 ppm in 2005. Levels of CO_2 are anticipated to reach between 600 and 1,000 ppm within the next one hundred years.

The science of climate change can be thought of as a movie that has been made by hundreds of directors and that takes viewers from billions of years ago to the present. But it is a film with many blurry images and empty frames. The goal of research in climate science is to refine these images and fill in the missing frames. As the movie chronicles actions that have brought us to the current era of global climate change, the challenge is to decipher the trajectory of the story in time to avoid a disastrous ending. The great majority of experts in the field believe that current evidence reliably indicates that anthropogenic causes have contributed to climate change. The movie, however, is unfinished. It leaves viewers wondering how the story will end.

Anticipating the outcome of the story is difficult. Prediction of the future is constrained by the limitations of scientific knowledge, the complexity of the climate system, uncertain population growth, an interconnected global economy, advances in technology, and changes in politics and policies. Given these uncertainties, conclusions that scientists reach on future changes in climate are expressed in terms of probabilities.

The problem for the international community is how to use uncertain predictions to decide what actions, if any, we should take to protect our world. Because global climate change is an environmental problem that has technological, social, and economic dimensions, international organizations, nations, businesses, and citizens must determine how they can limit their own contributions to harmful consequences. At least part of the world's reaction will be influenced by an understanding of how protection of the environment, the advancement of technology, and the development of our economy interconnect in conflicting ways.

References

Alley, Richard et al. 2007. *Climate change 2007: The physical science basis: Summary for policymakers*. Intergovernmental Panel on Climate Change. Geneva: IPCC Secretariat. ⟨www.ipcc.ch/SPM2feb07.pdf⟩ (accessed March 31, 2007).

Arrhenius, Svante. 1896. On the influence of carbonic acid in the air upon the temperature of the ground. *London, Edinburgh, and Dublin Philosophical Magazine and Journal of Science* (5th ser. April) 41: 237–75.

Arrhenius, Svante. 1908. *Worlds in the making: Evolution of the universe*. Translated by H. Borns. New York: Harper.

Calvin, William H. 2002. *A brain for all seasons: Human evolution and abrupt climate change*. Chicago: University of Chicago Press.

Crawford, Elisabeth. 1996. From ionic theory to the greenhouse effect. Canton, MA: Watson Publishing–Science History.

Davidson, Sarah. 2004. How global warming can chill the planet. *LiveScience*, 17 December. ⟨http://.livescience.com/forcesofnature/041217_sealevel_rise.html⟩ (accessed December 7, 2006).

Energy Information Administration. 2002. *Emissions of Greenhouse gases in the United States, 2001*. Washington, DC: EIA.

Environmental Protection Agency. 2006. The U.S. inventory of greenhouse gas emissions and sinks. 1990–2004. USEPA # 430-R-06-002 April, Washington, DC. ⟨http://www.epa.gov/globalwarming/publications/emissions⟩ (accessed December 2, 2006).

Fleming, James Rodger. 1998. *Historical perspectives on climate change*. New York: Oxford University Press.

Fleming, James Rodger. 2007. *The Callendar effect: The life and work of Guy Stewart Callendar*. Boston: American Meteorological Society.

Greenwood, David R., and Scott L. Wing. 1995. Eocene continental climates and latitudinal temperature gradients. *Geology* 23: 1044–48.

Karl, Thomas R., and Kevin E. Trenberth. 2003. Modern global climate change. *Science* 302(5651): 1719–23.

McCarthy, James J., Osvaldo F. Canziani, Neil A. Leary, David J. Dokken, and Kasey S. White, eds. 2001. *Climate change 2001: Impacts, adaptation and vulnerability.* Intergovernmental Panel on Climate Change. Cambridge: Cambridge University Press.

Met Office of the United Kingdom. 1997. *Climate change and its impacts: A global perspective.* Exeter/Devon: Met Office of the United Kingdom. ⟨http://www.metoffice.com/research/hadleycentre/pubs/brochures/B1997/index.html⟩ (accessed June 23, 2006).

National Academy of Sciences, Committee on the Science of Climate Change. 2001. *Climate change science: An analysis of some key questions.* Washington, DC: National Academies Press.

National Oceanic and Atmospheric Administration. 1994. *El Niño and climate prediction.* Reports to the Nation on Our Changing Planet. Boulder, CO: University Corporation for Atmospheric Research.

National Oceanic and Atmospheric Administration. 2006. Trends in atmospheric carbon dioxide, Boulder, CO. ⟨http://www.cmdl.noaa.gov/ccgg/trends⟩ (accessed December 2, 2006).

Robock A. 2002. Volcanic eruptions. In T. Munn, ed. *Encyclopedia of global environmental change.* Vol. 1. London: Wiley.

Sarmiento, J. L., and N. Gruber. 2002. Sinks for anthropogenic carbon. *Physics Today* 55(8): 30–36.

Seager, R. 2006. The source of Europe's mild climate. *American Scientist* 94: 334–41.

Severinghaus, J. P., T. Sowers, E. J. Brook, R. B. Alley, and M. L. Bender. 1998. Eocene continental climates and latitudinal temperature gradients. *Nature* 391: 141–46.

Shein, K. A. 2005. State of the climate in 2005. *Bulletin of the American Meteorological Society* 87(6): s1–s102.

Siegenthaler, U., T. Stocker, E. Monnin, D. Lüthi, J. Schwander, B. Stauffer, D. Raynaud, J. Barnola, H. Fischer, V. Masson-Delmotte, and J. Jouzel. 2005. Stable carbon cycle–climate relationship during the late Pleistocene. *Science* 310(5752): 1313–17.

World Meteorological Organization. 2002. *Executive Summary: Scientific Assessment of Ozone Depletion.* Global Ozone Research and Monitoring Project Report No. 47.

Zimov, Sergey A., Edward A. G. Schuur, and F. Stuart Chapin III. 2006. Permafrost and the global carbon budget. *Science* 312(5780): 1612–13.

3

Climate-Change Effects: Global and Local Views

John Abatzoglou, Joseph F. C. DiMento, Pamela Doughman, and Stefano Nespor

Climate-modeling studies have led to considerable knowledge about the effects of climate change at the worldwide, regional, and local levels. In this chapter, we give a broad overview of the effects of global climate changes and then focus on climate-change effects in a specific region: California.

Overall Global Effects of Climate Change

The scientific consensus on current climate change is that average global surface temperatures have risen 0.74°C (1.33°F) in the past century, with "most of the warming observed over the last fifty years" very likely due to the observed increase in anthropogenic greenhouse gas concentrations (Alley et al. 2007, 8). According to the World Meteorological Organization, the earth's ten warmest years on record occurred after 1990, and the warmest years in the observational record are 1998 and 2005. The observed warming over the past fifty years far exceeds any fifty-year period over the last one thousand years, with the warming rate accelerating in the 1980s and 1990s to 0.2°C (0.35°F) per decade. The National Academy of Science's 2001 report to the White House said that the global mean surface air temperature warmed in the twentieth century 0.4° to 0.8°C (0.7° to 1.5°F); that the oceans have warmed by about

0.05°C (0.09°F) over the water layer extending down 10,000 feet since the 1950s; and that this warming occurred mainly in noncontiguous ways throughout the century. But the report also said that the atmosphere at altitudes of about 13 miles has cooled in the last thirty-five years. In addition, the oceans are warming. Observations show that the uppermost 300 meters (1,000 feet) of the ocean have warmed 0.31°C (0.67°F) since the 1950s (Levitus et al. 2000).

With carbon dioxide levels already at 385 parts per million (over 100 ppm higher today than they were before the industrial revolution), experts agree that levels will continue to rise to between 600 ppm and 1,000 ppm by 2100. In its Fourth Assessment in 2007, the Intergovernmental Panel on Climate Change (IPCC) predicted an increase in the global average surface temperature of 1.8° to 4.0°C (3.24° to 7.2°F) by 2100 on the basis of its computer models of global climate (Alley et al. 2007). To put this in perspective, current global average surface temperatures are about 5°C (9°F) warmer than the most recent ice age. Other recent studies predict much higher global temperatures.[1] Even the low end of the estimated increases should rule out the "negligible warming" of natural climate variability that is proposed by some global-warming contrarians.

The notion of global warming as an enhanced greenhouse effect is best understood by using the following analogy. You try to stay warm at night by using a blanket to trap the heat that radiates away from your body. The blanket acts much like the earth's greenhouse gases do by reradiating heat back to the body of interest—to you in the former and to the earth's surface in the latter. Just as a blanket must be the right thickness to keep you comfortable, the earth's greenhouse gases must be of the right type to keep the planet comfortable. And just as too many blankets may make you too warm, too many greenhouse gases may lead to a warming of the globe.

Significant increases in global mean temperature are projected to result in heightened health risks for much of the world's population (Adger et al. 2007). Heat-related mortality increases significantly on days when the air temperature tops 32°C (90°F) (Davis et al. 2003). By raising the baseline average temperature, heat waves will become more intense, and the number of days that exceed a given temperature will also grow. These changes will increase the number of heat-related deaths, especially among the elderly and the urban poor. The likelihood of heat waves, such as those that ravaged much of Europe in the summer of 2003 and resulted in over thirty thousand cases of heat prostration, is expected to increase a hundredfold over the next forty years as a result of anthropogenic climate change (Stott et al. 2004). Scientists agree that the summer of 2003 was likely the warmest summer recorded in Europe over the past five hundred years and noted that increases in global temperature drastically increased the probability of the heat wave. Ironically, what today is classified as an extremely warm season (like the summer of 2003 across much of Europe) is expected to be classified as an extremely cool season by the end of the twenty-first century.

Each year, extreme weather events such as floods and heat waves are responsible for the loss of thousands of human lives, billions of dollars in damages, and irreversible environmental harms. Extreme events are the main channel where climate and social and economic systems interact, and they attract the most climate-related media reports. Ecosystems are able to cope with so-called normal climate conditions, but both ecosystems and human beings are vulnerable to extreme events such as droughts, heat waves, and floods. Thus, the harshest effects of climate change in the twenty-first century will be experienced through changes in extreme events. Extreme weather events contributed to over $200 billion of damages worldwide in 2005 (Munich Re Group 2005).

Changes in extreme weather events are already apparent. The Intergovernmental Panel on Climate Change reported an increase in heat-wave frequency, intensity, and duration and a 2 to 4 percent increase in the number of heavy-precipitation events across the Northern Hemisphere during the twentieth century. However, present changes are benign compared to the anticipated changes in extreme events during the twenty-first century. Modeling studies have suggested that there will be a significant increase in the frequency of extreme precipitation events and heat waves in an enhanced greenhouse climate (Meehl, Arblaster, and Tebaldi 2005).

Extreme weather events (including tropical storms) have been at the forefront of media attention after the the summer of 2005. Tropical-storm records for the Atlantic basin dropped like dominoes during 2005. There were twenty-seven named storms (old record 21). Fifteen of these storms became hurricanes (a tropical storm with winds exceeding 74 mph) (old record 12), and four attained the most powerful category 5 status (winds exceeding 155 mph), including Hurricane Wilma, the most powerful hurricane ever recorded in the Atlantic (with winds exceeding 175 mph) (old record 2). Several of these powerful hurricanes (most notably Hurricane Katrina) made landfall across the United States, Caribbean, and Central America, resulting in over 2,500 fatalities and over $100 billion in damage. (In the 2006 season there were only nine named storms.)

No one hurricane (such as Katrina) or one storm season (such as the record-breaking 2005 tropical storm season in the

Atlantic) can be attributed to climate change, but scientists are connecting increases in tropical-storm intensity to warmer surface ocean temperatures. Warm surface ocean temperatures in the tropics provide the energy that fuels tropical storms, so a warming of the tropical ocean would be expected to lead to more powerful tropical storms. Theoretical results confirm these expectations by suggesting that tropical-storm wind speeds increase 5 percent for every 1°C (1.8°F) increase in surface ocean temperatures (Emanuel 1987). Winds from tropical storms are devastating in their own regard and serve as a catalyst in producing storm surges (which are typically associated with the largest number of casualties).

Although there appears to be no global trend in tropical-storm frequency, the number of major hurricanes (categories 4 and 5) has nearly doubled in the last thirty-five years (Webster et al. 2005). This observation is consistent with the 0.8° to 2.4°C (1.4° to 4.3°F) increase in tropical surface ocean temperatures over the last fifty years. Additional evidence shows that maximum wind speeds in tropical storms have increased 15 percent over the last thirty years (Emanuel 2005). These results suggest that tropical storms are getting stronger, but some members of the scientific community insist that a longer historical record of tropical storms is needed before global warming can be pointed to as the cause. Tropical surface ocean temperatures are expected to increase 1° to 2.5°C (1.8° to 4.5°F) over the next one hundred years, and state-of-the-art numerical models suggest that warmer waters will beget stronger and more devastating tropical storms in the coming century (Knutsen and Tuleya 2004).

Global warming's effects extend across all sectors of health, the economy, politics, and international relations. Increased humidity worsens urban air pollution (McCarthy et al. 2001, 12); malaria and dengue are spread by vector-borne infections (and mosquito populations increase with moisture); energy sources and water resources are stressed by storms; storms damage agricultural sectors and threaten food supplies; and potentially higher sea levels would displace millions of "climate refugees."

Natural ecosystems in the industrialized world also face the effects of climate change. Climate change may result in the loss of biodiversity, may increase extinction rates for vulnerable species, and may cause a decline in the viability of important ecosystems (McCarthy et al. 2001, 5). Sea-level rise and the warming of the ocean will further harm coral reefs, which provide the greatest biodiversity of any marine ecosystem. The growing seasons in temperate regions will lengthen, and plant and animal ranges will shift poleward and move to higher elevations. Signs of spring will shift as well: trees will flower, insects will emerge, and birds will lay eggs earlier in the year. This is already happening: the spring bloom in the northeastern United States is arriving four to eight days earlier today than it did during the 1960s (Clean Air, Cool Planet 2006). Plant and animal species currently labeled as critically endangered will become extinct, and the majority of those labeled as "endangered or vulnerable" will come closer to extinction (McCarthy et al. 2001, 11). Given a slow enough change in climate, many species will be able to adapt to the changes, but rapid changes in climate severely limit adaptation strategies.

The IPCC predicts that the global average sea level will rise 0.28 to 0.43 meters (0.92 to 1.40 feet) by 2100 (Alley et al. 2007). A global average increase of 0.1 to 0.2 meters (0.32 to 0.65 feet) was observed for the twentieth century (Houghton et al. 2001a, 21). Over the past fifteen years, the rate of global sea-level rise has been about 0.03 meters per decade (0.1 foot per decade) (Leuliette, Nerem, and Mitchum 2004). The bulk of the increase in sea-level rise today is a result of warmer temperatures. Because of thermal expansion, increases in oceanic temperature result in an increased volume of the ocean without the addition of any mass. Computer models predict that the thickness of Arctic sea ice will decline, the Greenland ice sheet will melt, and the area of ice-free seas will increase. In 2005, the extent of sea ice in the Northern Hemisphere set record minimum values in eleven of twelve months (Shein 2005). This observation is consistent with the present decreasing trend of sea ice of 2 percent per decade during early spring and 7 percent per decade during early fall (Stroeve et al. 2005). Although melting sea ice will not directly contribute to sea-level rise, the melting of large continental glaciers, such as those in Greenland and Antarctica, would add a huge amount of mass to the oceans. In addition, a majority of the world's high-altitude glaciers are currently eroding. For example, the glaciers of Glacier National Park in Montana are expected to melt within the next thirty years (Hall and Fagre 2003).

Because the ocean has a large heat capacity, it takes a long time to translate changes in atmospheric temperature to changes in oceanic temperature. Researchers conclude that even if the atmosphere stopped warming today, the global sea level would continue to rise for centuries before reaching a new equilibrium (Houghton et al. 2001a, 16). The IPCC models suggest a total rise in sea level of up to 7 meters (23 feet) by the end of the millennium if the Greenland ice sheet is eliminated. Sea-level rise is expected to cause "increased levels of flooding, accelerated erosion, loss of wetlands and mangroves and seawater intrusion into freshwater sources" (Houghton et al. 2001a). Many of the people who live near present-day sea level will likely be displaced as waters inundate the land.

Climate scientists also expect that the *hydrologic cycle*—the cycling of water among the atmosphere, land, and oceans through precipitation and evaporation—will intensify because of global warming. A warmer climate will lead to enhanced rates of evaporation and increased precipitation for the globe as a whole (Wetherald and Manabe 2002). Regional precipitation distributions may be drastically altered, leading to an increase in the intensity and frequency of rainfall in some regions and of droughtlike conditions in other regions. With depleting snow packs, rising sea levels, and increasingly severe deluges, floods will likely increase in number and severity. It is also generally believed that in a warmer world the atmospheric concentration of water vapor will increase and that increased water vapor (which is an important greenhouse gas) may further enhance the predicted warming.

More Regional Effects

In addition to changes in global average climate models predict substantial regional differences in climate responses. For some elements, such as temperature or precipitation, large regional changes may occur that oppose the global mean change (McCarthy et al. 2001, 938). For example, some scientists expect decreases in temperature across the Southern Ocean and Antarctica, despite the overall warming of the planet.

Regional impacts of climate change are extensive. Coastal settlements across much of Africa will experience erosion and inundation (McCarthy et al. 2001, 3). In Asia, tens of millions of people in low-lying coastal areas will be displaced by rising sea levels, while mangroves and coral reefs will be at risk (McCarthy et al. 2001, 14). Sea-level rise will also likely consume the last remaining habitat of the Bengal tiger (McCarthy et al. 2001, 929). Species in Australia and New Zealand that thrive in restricted areas (called *climactic niches*) will become endangered or extinct (McCarthy et al. 2001, 15). The rising ocean will also consume many Pacific island countries. In Latin America, home to some of the largest concentrations of biodiversity, climate change may result in the eradication of countless vulnerable species. Agricultural productivity in Latin America and Africa will decline, leading to food shortages for many impoverished countries. Recent work by the European Environment Agency suggests that changes in temperature across much of Europe will exceed the global mean warming. The Agency also foresees wetter and much warmer conditions for northern Europe and persistent droughtlike conditions with deleterious effects on agriculture in southern Europe.

Polar areas will see the most rapid and largest changes in climate over the next century. Inuit communities in the central Canadian Arctic, whose culture and lives revolve around ice, have reported significant increases in temperature and thawing ice over the last thirty years. These changes have resulted in a declining food supply, an invasion of nonindigenous species, and threats to a life style that depends on permafrost. A report released by the Arctic Council (2004) confirmed scientific expectations that the Arctic has warmed more rapidly than other regions. It found that the coverage of Arctic sea ice has declined as much as 30 percent in the last fifty years, that coastal Alaskan villages are relocating inland, that animals dependent on sea ice are in decline and some are becoming threatened species, and that the West Nile virus is moving into the northern provinces of Canada. In fact, global warming appears to have arrived at northern latitudes. On average, temperatures in the Arctic region have increased 1.2°C (2.1°F) over the last century. Locally, temperatures in Russia and parts of Alaska have risen 6°C (10.8°F) since the 1970s and are at their highest in four hundred years (McFarling 2004).

Yet global warming over some of the coldest places on the earth is not necessarily bad news for all life in these areas. The potential benefits of climate change at higher latitudes include growth in stocks of marine fish and improved prospects for agriculture and timber harvests (Revkin 2004).

To understand the sensitivity of climate change in the context of Arctic sea ice, consider the following. Warming melts sea ice, and as bright, highly reflective sea ice melts away, the dark, highly absorptive ocean surface is exposed. Instead of solar radiation being reflected back to space from bright sea ice, it is instead absorbed by the planet, augmenting the earth's radiation budget and warming the planet. This leads to further sea ice melt and so on. This positive feedback amplifies the anticipated global warming and implies that polar regions will warm up significantly more than the rest of the globe. Current decreasing trends of Arctic sea ice suggest that this feedback mechanism is presently at play.

In North America, in some mountainous areas, warmer conditions will decrease the amount of precipitation that falls in the form of snow, thereby leading to earlier peak flows of water and earlier growing seasons. As nearly 75 percent of the water supply for the western United States resides in mountain snow packs, water shortages could increase toward the end of summer and fall as the climate warms. Increased evaporation rates associated with warmer temperatures may also increase the likelihood of long droughts. Furthermore, these conditions are likely to increase the frequency and intensity of large wildfires, such as those experienced in recent years across much of the Pacific Northwest, California, Alaska, Colorado, Arizona, Texas, and Oklahoma (Brown, Hall, and Westerling 2004).

In the Midwestern United States, some crops may benefit from climate change—a result of carbon dioxide fertilization and favorable precipitation and temperature conditions. However, warmer temperatures will push agriculture northward across the continent, and prime cropland over the southern Great Plains of the United States will be much less productive and profitable. In New England, decorative maple trees will be replaced by species such as oak that are tolerant to warmer summer conditions. Farther south, in addition to the already noted increase in tropical-storm intensity, rising sea levels will amplify the risk of damaging storm surges along much of the Atlantic and Gulf coasts (McCarthy et al. 2001, 15). In much of low-lying Florida, sea-level rise alone will allow waters to advance up to 400 feet inland, flooding shoreline infrastructure and disrupting freshwater supplies. Finally, many major cities in the United States, including New York and Chicago, are expected to see drastic increases in the frequency and intensity of heat waves over the next one hundred years. For example, it is anticipated that by the year 2100 New York will exceed the 32°C (90°F) threshold as frequently as present-day Houston does (Bloomfield, Smith, and Thompson 1999).

A Close Look: California

To examine global climate change at the local level, we turn to the observed changes and future forecasts in climate for California. A recent paper by the National Academy of Sciences describes climate change for the state of California as providing "a challenging test case to evaluate impacts of regional-scale climate change" because of the complex interactions among the diverse climate zones statewide. California is also the most populous U.S. state and contributes about 6 percent of total U.S. emissions of greenhouse gases. Its environmental policies often are harbingers of what other states and regions will adopt, and it has taken the lead on many policies that are linked to regulation of and adaptation to climate change.

The best available scientific conclusions paint the following picture of California over the next several decades. As in most of the earth's regions, the average temperature in California has increased over the last century. For example, Fresno (located in the central valley of the state) has seen its average annual temperature increase from an average of 16°C (60.9°F) in 1899 to 1928 to an average of 17.6°C (63.7°F) in 1971 to 2000. Following the observed trend, climate experts say that temperatures will continue to rise over the next century.

As is the case for the globe as a whole, temperatures are expected to warm much more dramatically than the present observed rate of change. Over the past half century statewide temperatures have increased 0.9°C (1.6°F), with the warming trends for minimum temperatures increasing at twice the rate of maximum temperatures. Results from a large number of modeling runs project that statewide temperatures will increase 2 to 7°C (3.6° to 12.6°F) over the course of the twenty-first century (Dettinger 2005). A scientist described the way this would affect Californians: it is "enough to make many coastal cities feel like inland cities do today, and enough to make inland cities feel like Death Valley" (Hayhoe et al. 2004).

An increase in average temperatures means that heat waves will be more frequent and more intense. In Los Angeles, the number of heat-wave days (defined as a period of at least three days in a row with maximum temperatures exceeding 32°C (90°F) is expected to increase eightfold, and heat-related deaths are expected to increase. During July 2006, an extremely warm air mass stagnated over the western United States. Fresno experienced six consecutive days of 43°C (110°F) temperatures; the low temperature in Death Valley on July 24 was 38°C (100°F). An extended period of record-setting heat during the latter half of the month contributed to the death of over 160 people statewide (Steinhauer 2006). Although it is too early to attribute the heatwave of 2006 to global warming, models predict that extended warm periods will be the norm rather than the exception.

Scientists also predict changes in both the quantity and form of precipitation. Precipitation totals are expected to increase 20 to 30 percent, although these predictions are susceptible to a wide range of uncertainty, and some studies predict a slight decrease in precipitation in the second half of the century (Dettinger 2005). Of greater importance to the water-deprived urban and agricultural sections of the state is that warmer temperatures will result in more precipitation falling in the form of rain rather than snow. This means that less water will be stored in the mountains in the snow pack (by far the state's largest reservoir). Since the rainy season in California is confined primarily to winter, the state relies heavily on snow melt for water during the dry months. The snow pack will decline up to 70 percent, mainly at lower elevations (Hayhoe et al. 2004). As a result, the once reliable source of water during the dry season will be depleted, and water shortages could be the rule (King et al. 1989).

In California, water runoff is expected to decline for much of the spring, summer and fall. But warmer and wetter winter storms with more rain than snow will result in dramatically increased winter runoff. Since reservoirs cannot store anticipated increases in winter runoff (if in the form of rain only) for use during the dry season, flooding will likely be more frequent (Miller et al. 2003). The increased flood risk will probably be exacerbated by higher sea levels. Sea level is currently rising by 0.1 to 0.2 meters (0.32 to 0.65 feet) per century along much of the state's coast. With the accelerated pace of global warming, sea levels are likely to rise another 0.3 to 0.5 meters (0.95 to 1.6 feet) by 2100 (EPA 1997). Increased flooding and rising sea level may upset the delicate balance of saltwater and freshwater inflows into the Sacramento–San Joaquin delta. Levee systems in the delta are already vulnerable, so added pressures are likely to increase the risk of intrusion of salt water into freshwater areas. This could further threaten the state's water supply.

Any decline in water availability would damage California's multibillion dollar agriculture industry. A combination of more winter floods and a drier growing season means that water-intensive crops are likely to suffer (Field et al. 1999; Wilkinson et al. 2002; Miller, Bashford, and Strem 2003). Warmer temperatures and decreased soil moisture would dramatically increase irrigation demands, and total yields from crops such as cotton and wheat could decline between 15 to 45 percent (EPA 1997). The combination of a longer growing season and augmented carbon dioxide fertilization, however, may actually increase the yields of hay, citrus, and tomato crops (EPA 1997). The quality and quantity of wines harvested from the cool coastal regions of California are actually expected to improve with the predicted changes, but many of the inland and valley regions will experience changes in climate that would make them inhospitable to fine wine production (Hayhoe et al. 2004; Nemani et al. 2004).

Urban regions of California that currently experience periods of very hot and dry conditions are likely to see an increase in heat-related illnesses and deaths (EPA 1997). Despite recent improvements in air quality over parts of the state, air quality is predicted to degrade as ground-level ozone increases. The area of the state that fails to meet air-quality standards is predicted to expand greatly over the next one hundred years, based on estimates that population and industry will continue to grow and spread and that air pollution worsens during hot weather (EPA 1997). The U.S. EPA (1997) offers the following estimate of the magnitude of the problem:

In the Bay Area and the Central Valley, with no other changes in weather or emissions, a 7.2°F (4°C) warming would increase ozone concentrations by 20 percent and almost double the size of the area not meeting national health standards for air quality. Ground-level ozone has been shown to aggravate existing respiratory illnesses such as asthma, reduce lung function, and induce respiratory inflammation.

Predicted statewide increases in temperature and population will increase the demand for air conditioning. The peak statewide demand for electricity in California in 2002 was over 51,800 megawatts during the hottest hour of the year. In contrast, the demand was less than 38,600 megawatts for 90 percent of the year (Wetherall 2004). Meeting peak electricity demand requires a sizeable investment in power plants, many of which are utilized for only a few hours a year to meet spikes in electricity demand during extremely hot days. With expected increases in population and temperature, peak electricity demands will skyrocket, and more power plants will be needed.

Scientists also expect changes in the survival and distribution of certain species of plants and animals within the state (Field et al. 1999). Climate change in California is expected to alter the timing of important ecological events, such as the onset of the bloom, and also to shift ecological zones northward and upward. Species migration and adaptation will likely decrease biodiversity. Certain species may have problems adapting to climate change. For example, if butterflies emerge before the flowers they depend on, populations will decline. If ecological niches are lost, the extinction of species such as the bighorn sheep and the rosy finch is possible. Even with the modest warming experienced over the twentieth century, Edith's checkerspot butterfly is disappearing from its southernmost range in California. The productivity of coniferous forests is expected to decrease while the incidence of "disturbance factors" such as fires and outbreaks of insect infestations will likely increase (King et al. 1989; McCarthy et al. 2001).

Longer and more intense summers and drier conditions are likely to increase wildfires across the state (Brown, Hall, and Westerling 2004). To maintain ideal growing conditions, vegetation will also shift northward. For example, a 3°C (5.4°F) warming over the next century would require plants and forests to migrate northward about 5 kilometers per year. This may not be plausible; it is more likely that alpine forests will be replaced by grassland and chaparral throughout California (EPA 1997).

Conclusion

Climate change was recognized as a global environmental problem in the late 1970s, spurring an increase in international scientific efforts to understand its causes and effects. Climate models have been developed that allow researchers to forecast climate well into the future. By adjusting for population and economic changes and accounting for the interactions that comprise the climate system, researchers have been able to grasp what the climate of the future holds. Global average temperatures are expected to rise 1.8° to 4.0°C (3.24° to 7.2°F) over the next century. Regional impacts will vary widely. Some changes have already begun, while others are expected to occur over the next fifty to one hundred years. The rate of climate change is a crucial component in how the planet weathers the storm. If warming is slow, adaptation and mitigation may limit damages, but abrupt climate change would likely have severe consequences.

We have a problem with no easy solution. Although carbon is a building block for life on earth, carbon dioxide remains in the atmosphere for about 120 years. Changes have been set in motion. What are we going to do about it?

Notes

1. One published in *Nature* concluded that temperatures could increase up to 11°C (19°F). The chief scientist on the study stated, "An 11°C-warmed world would be a dramatically different world.... There would be large areas at higher latitudes that could be up to 20°C (36°F) warmer than today. The UK would be at the high end of these changes. It is possible that even present levels of greenhouse gases maintained for long periods may lead to dangerous climate change.... When you start to look at these temperatures, I get very worried indeed" (Stainforth et al. 2005, 404).

References

Adger, Neil et al. 2007. *Climate change 2007: Climate change impacts, adaptation, and vulnerability: Summary for policymakers.* Intergovernmental Panel on Climate Change. Brussels: IPCC. ⟨http://www.ipcc.ch⟩ (accessed April 7, 2007).

Alley, Richard et al. 2007. *Climate change 2007: The physical science basis: Summary for policymakers.* Intergovernmental Panel on Climate Change. Geneva: IPCC Secretariat. ⟨http://www.ipcc.ch/SPM2feb07.pdf⟩ (accessed March 31, 2007).

Arctic Council. 2004. *Arctic climate impact assessment.* Arctic Council, Oslo, Norway ⟨http://.acia.uaf.edu⟩ (accessed March 14, 2005).

Bloomfield, Janine, M. Smith, and N. Thompson. 1999. *Hot nights in the city: Global warming, sea-level rise and the New York metropolitan region.* New York: Environmental Defense Fund. ⟨http://www.environmentaldefense.org/documents/493_HotNY.pdf⟩ (accessed December 2, 2006).

Brown, T. J., B. L. Hall, and A. L. Westerling. 2004. The impact of twenty-first century climate change on wildland fire danger in the western United States: An applications perspective. *Climatic Change* 62: 365–88.

Clean Air, Cool Planet. 2006. Evidence of early spring indicators of climate change for the Northeast. Portsmouth, NH ⟨http://www.cleanair-coolplanet.org⟩ (accessed December 2, 2006).

Davis, R. E., P. C. Knappenberger, W. M. Novicoff, and P. J. Michaels. 2003. Decadal changes in summer mortality in U.S. cities. *International Journal of Biometeorology* 47: 166–75.

Dettinger, M. D. 2005. From climate-change spaghetti to climate change distributions for 21st century California. *San Francisco Estuary and Watershed Science* 3, no. 1 (March), article 4.

Emanuel, K. A. 1987. The dependence of hurricane intensity on climate. *Nature* 326: 483–85.

Emanuel, K. A. 2005. Increasing destructiveness of tropical cyclones over the past thirty years. *Nature* 436: 686–88. ⟨http://.nature.com/nature/journal/v436/n7051/full/nature03906.html⟩ (accessed December 2, 2006).

Environmental Protection Agency (EPA). 1997. Climate change and California. EPA, Washington, DC. ⟨http://yosemite.epa.gov/oar/globalwarming.nsf/UniqueKeyLookup/SHSU5BNNKT/$File/ca_impct.pdf⟩ (accessed June 21, 2005).

Field, C. B., G. C. Daily, F. W. Davis, S. Gaines, P. A. Matson, J. Melack, and N. L. Miller. 1999. *Confronting climate change in California: Ecological impacts on the Golden State.* Cambridge, MA: Union of Concerned Scientists; Washington, DC: Ecological Society of America.

Hall, Myrna H. P., and Daniel B. Fagre. 2003. Modeled climate-induced glacier change in Glacier National Park, 1850–2100. *BioScience* 53: 131–40.

Hayhoe, Katharine, D. Cayan, C. B. Field, P. C. Frumhoff, E. P. Maurer, N. L. Miller, S. C. Moser, S. H. Schneider, K. N. Cahill, E. E. Cleland, L. Dale, R. Drapek, R. M. Hanemann, L. S. Kalkstein, J. Lenihan, Claire K. Lunch, Ronald P. Neilson, S. C. Sheridan, and Julia H. Verville. 2004. Emission pathways, climate change, and impacts on California. *Proceedings of the National Academy of Sciences* 101: 12422–27.

Houghton, J. T., Y. Ding, D. J. Griggs, M. Noguer, P. J. van der Linden, X. Dai, K. Maskell, and C. A. Johnson, eds. 2001a. *Climate change 2001: The scientific basis.* Intergovernmental Panel on Climate Change. Cambridge: Cambridge University Press.

Houghton, J. T., Y. Ding, D. J. Griggs, M. Noguer, P. J. van der Linden, X. Dai, K. Maskell, and C. A. Johnson, eds. 2001b. *Climate change 2001: The scientific basis* (Third Assessment Report). Intergovernmental Panel on Climate Change (IPCC). Cambridge: Cambridge University Press.

King, G. A., R. L. DeVelice, R. P. Neilson, and R. C. Worrest. 1989. California. In *The potential effects of global climate change on the United States*. Report EPA-230-05-89-050. Washington, DC: U.S. Environmental Protection Agency.

Knutson, T. R., and R. E. Tuleya. 2004. Impact of CO_2-induced warming on simulated hurricane intensity and precipitation: Sensitivity to the choice of climate model and convective parameterization. *Journal of Climate* 17: 3477–95.

Leuliette, E. W., R. S. Nerem, and G. T. Mitchum. 2004. Calibration of TOPEX/*Poseidon* and *Jason* altimeter data to construct a continuous record of mean sea level change. *Marine Geodesy* 27: 79–94. ⟨http://sealevel.colorado.edu⟩ (accessed December 2, 2006).

Levitus, S., J. I. Antonov, T. P. Boyer, and C. Stephens. 2000. Warming of the world ocean. *Science* 287: 2225–29.

McCarthy, James J., Osvaldo F. Canziani, Neil A. Leary, David J. Dokken, and Kasey S. White, eds. 2001. *Climate change 2001: Impacts, adaptation and vulnerability*. Intergovernmental Panel on Climate Change. Cambridge: Cambridge University Press.

McFarling, Usha Lee. 2004. Climate change accelerating, report warns. *Los Angeles Times*, November 9, A1.

Meehl, G. A., J. M. Arblaster, and C. Tebaldi. 2005. Understanding future patterns of precipitation extremes in climate model simulations. *Geophysical Research Letters* 32: L18719.

Miller, N. L., K. E. Bashford, and E. Strem. 2003. Potential impacts of climate change on California hydrology. *Journal of the American Water Resources Association* 39: 771–84.

Munich Re Group. 2005. Press release, December 29, 2005. ⟨http://www.munichre.com/en/press/press_releases/2005/2005_12_29_press_release.aspx⟩ (accessed April 3, 2007).

National Academy of Sciences, Committee on the Science of Climate Change. 2001. *Climate change science: An analysis of some key questions*. Washington, DC: National Academies Press.

Nemani, R. R., M. A. White, D. R. Cayan, G. V. Jones, S. W. Running, J. C. Coughlan, and D. L. Peterson. 2004. Asymmetric warming over coastal California and its impact on the premium wine industry. *Climate Research* 19: 25–34.

Revkin, Andrew C. 2004. Big Arctic perils seen in warming. *New York Times*, October 30, A1.

Stainforth, D., T. Aina, C. Christensen, M. Collins, N. Faull, D. J. Frame, J. A. Kettleborough, S. Knight, A. Martin, J. M. Murphy, C. Piani, D. Sexton, L. A. Smith, R. A. Spicer, A. J. Thorpe, and M. R. Allen. 2005. Uncertainty in predictions of the climate response to rising levels of greenhouse gases. *Nature* 433: 403–06.

Shein, K. A. 2005. State of the climate in 2005, executive summary. *Bulletin of the American Meteorological Society* 87: 801–805. ⟨http://ams.allenpress.com/perlserv/?request=get-abstract&doi=10.1175/BAMS-87-6-shein⟩ (accessed April 3, 2007).

Steinhauer, Jennifer. 2006. In California, heat is blamed for one hundred deaths. *New York Times*, July 27, A24.

Stott, Peter A., D. A. Stone, and M. R. Allen. 2004. Human contribution to the European heatwave of 2003. *Nature* 432: 610–14.

Stroeve, J., C. F. Fetterer, K. Knowles, W. Meier, M. Serreze, and T. Arbetter. 2005. Tracking the Arctic's shrinking ice cover: Another extreme September minimum in 2004. *Geophysical Research Letters* 32: L04S01.

Webster, P. J., G. J. Holland, J. A. Curry, and H. R. Chang, 2005. Changes in tropical cyclone number, duration, and intensity in a warming environment. *Science* 309: 1844–46.

Wetherald, Richard T., and Syukuro Manabe. 2002. Simulation of hydrologic changes associated with global warming. *Journal of Geophysical Research* 107: 4379.

Wetherall, Ron. 2004. *CEC environmental performance report: California electricity system overview*. California Energy Commission, November 15. ⟨http://.energy.ca.gov/2005_energypolicy/documents/2004-11-15_workshop/2004-11-15_02-B_wetherall.pdf⟩ (accessed June 23, 2006).

Wilkinson, R., K. Clarke, M. Goodchild, J. Reichman, and J. Dozier. 2002. *The potential consequences of climate variability and change for California: The California regional assessment*. Washington, DC: U.S. Global Change Research Program.

4

The Scientific Consensus on Climate Change: How Do We Know We're Not Wrong?

Naomi Oreskes

In December 2004, *Discover* magazine ran an article on the top science stories of the year. One of these was climate change, and the story was the emergence of a scientific consensus over the reality of global warming. *National Geographic* similarly declared 2004 the year that global warming "got respect" (Roach 2004).

Many scientists felt that respect was overdue: as early as 1995, the Intergovernmental Panel on Climate Change (IPCC) had concluded that there was strong scientific evidence that human activities were affecting global climate. By 2007, the IPCC's Fourth Assessment Report noted it is "extremely unlikely that the global climate changes of the past fifty years can be explained without invoking human activities" (Alley et al. 2007). Prominent scientists and major scientific organizations have all ratified the IPCC conclusion. Today, all but a tiny handful of climate scientists are convinced that earth's climate is heating up and that human activities are a significant cause.

Yet many Americans continue to wonder. A recent poll reported in *Time* magazine (Americans see a climate problem 2006) found that only just over half (56 percent) of Americans think that average global temperatures have risen despite the fact that virtually all climate scientists think that they have.[1]

More startlingly, a majority of Americans believe that scientists are still divided about the issue. In some quarters, these doubts have been invoked to justify the American refusal to join the rest of the world in addressing the problem.

This book deals with the question of climate change and its future impacts, and by definition predictions are uncertain. People may wonder why we should spend time, effort, and money addressing a problem that may not affect us for years or decades to come. Several chapters in this book address that question—explaining how some harmful effects are already occurring, how we can assess the likely extent of future harms, and why it is reasonable to act now to prevent a worst-case scenario from coming true.

This chapter addresses a different question: might the scientific consensus be wrong? If the history of science teaches anything, it's humility. There are numerous historical examples where expert opinion turned out to be wrong. At the start of the twentieth century, Max Planck was advised not to go into physics because all the important questions had been answered, medical doctors prescribed arsenic for stomach ailments, and geophysicists were confident that continents could not drift. Moreover, in any scientific community there are always some individuals who depart from generally accepted views, and occasionally they turn out to be right. At present, there is a scientific consensus on global warming, but how do we know it's not wrong?

The Scientific Consensus on Climate Change

Let's start with a simple question: What is the scientific consensus on climate change, and how do we know it exists? Scientists do not vote on contested issues, and most scientific questions are far too complex to be answered by a simple yes or no, so how does anyone know what scientists think about global warming?

Scientists glean their colleagues' conclusions by reading their results in published scientific literature, listening to presentations at scientific conferences, and discussing data and ideas in the hallways of conference centers, university departments, research institutes, and government agencies. For outsiders, this information is difficult to access: scientific papers and conferences are by experts for experts and are difficult for outsiders to understand.

Climate science is a little different. Because of the political importance of the topic, scientists have been unusually motivated to explain their research results in accessible ways, and explicit statements of the state of scientific knowledge are easy to find.

An obvious place to start is the Intergovernmental Panel on Climate Change (IPCC), already discussed in previous chapters. Created in 1988 by the World Meteorological Organization and the United Nations Environment Program, the IPCC evaluates the state of climate science as a basis for informed policy action, primarily on the basis of peer-reviewed and published scientific literature (IPCC 2005). The IPCC has issued four assessments. Already in 2001, the IPCC had stated unequivocally that the consensus of scientific opinion is that earth's climate is being affected by human activities. This view is expressed throughout the report, but the clearest statement is: "Human activities . . . are modifying the concentration of atmospheric constituents . . . that absorb or scatter radiant energy. . . . [M]ost of the observed warming over the last 50 years is likely to have been due to the increase in greenhouse gas concentrations" (McCarthy et al. 2001, 21). The 2007

IPCC reports says "very likely" (Alley et al. 2007). The IPCC is an unusual scientific organization: it was created not to foster new research but to compile and assess existing knowledge on a politically charged issue. Perhaps its conclusions have been skewed by these political concerns, but the IPCC is by no means alone it its conclusions, and its results have been repeatedly ratified by other scientific organizations.

In the past several years, all of the major scientific bodies in the United States whose membership's expertise bears directly on the matter have issued reports or statements that confirm the IPCC conclusion. One is the National Academy of Sciences report, *Climate Change Science: An Analysis of Some Key Questions* (2001), which originated from a White House request. Here is how it opens: "Greenhouse gases are accumulating in Earth's atmosphere as a result of human activities, causing surface air temperatures and subsurface ocean temperatures to rise" (National Academy of Sciences 2001, 1). The report explicitly addresses whether the IPCC assessment is a fair summary of professional scientific thinking and answers yes: "The IPCC's conclusion that most of the observed warming of the last 50 years is likely to have been due to the increase in greenhouse gas concentrations accurately reflects the current thinking of the scientific community on this issue" (National Academy of Sciences 2001, 3).

Other U.S. scientific groups agree. In February 2003, the American Meteorological Society adopted the following statement on climate change: "There is now clear evidence that the mean annual temperature at the Earth's surface, averaged over the entire globe, has been increasing in the past 200 years. There is also clear evidence that the abundance of greenhouse gases has increased over the same period.... Because human activities are contributing to climate change, we have a collective responsibility to develop and undertake carefully considered response actions" (American Meteorological Society 2003). So too says the American Geophysical Union: "Scientific evidence strongly indicates that natural influences cannot explain the rapid increase in global near-surface temperatures observed during the second half of the 20th century" (American Geophysical Union Council 2003). Likewise the American Association for the Advancement of Science: "The world is warming up. Average temperatures are half a degree centigrade higher than a century ago. The nine warmest years this century have all occurred since 1980, and the 1990s were probably the warmest decade of the second millennium. Pollution from 'greenhouse gases' such as carbon dioxide (CO_2) and methane is at least partly to blame" (Harrison and Pearce 2000). Climate scientists agree that global warming is real and substantially attributable to human activities.

These kinds of reports and statements are drafted through a careful process involving many opportunities for comment, criticism, and revision, so it is unlikely that they would diverge greatly from the opinions of the societies' memberships. Nevertheless, it could be the case that they downplay dissenting opinions.[2]

One way to test that hypothesis is by analyzing the contents of published scientific papers, which contain the views that are considered sufficiently supported by evidence that they merit publication in expert journals. After all, any one can *say* anything, but not anyone can get research results published in a refereed journal.[3] Papers published in scientific journals must pass the scrutiny of critical, expert colleagues. They must be supported by sufficient evidence to convince others who know the subject well. So one must turn to the scientific literature to be certain of what scientists really think.

Before the twentieth century, this would have been a trivial task. The number of scientists directly involved in any given debate was usually small. A handful, a dozen, perhaps a hundred, at most, participated—in part because the total number of scientists in the world was very small (Price 1986). Moreover, because professional science was a limited activity, many scientists used language that was accessible to scientists in other disciplines as well as to serious amateurs. It was relatively easy for an educated person in the nineteenth or early twentieth century to read a scientific book or paper and understand what the scientist was trying to say. One did not have to be a scientist to read *The Principles of Geology* or *The Origin of Species*.

Our contemporary world is different. Today, hundreds of thousands of scientists publish over a million scientific papers each year.[4] The American Geophysical Union has 41,000 members in 130 countries, and the American Meteorological Society has 11,000. The IPCC reports involved the participation of many hundreds of scientists from scores of countries (Houghton, Jenkins, and Ephraums 1990; Alley et al. 2007). No individual could possibly read all the scientific papers on a subject without making a full-time career of it.

Fortunately, the growth of science has been accompanied by the growth of tools to manage scientific information. One of the most important of these is the database of the Institute for Scientific Information (ISI). In its Web of Science, the ISI indexes all papers published in refereed scientific journals every year—over 8,500 journals. Using a key word or phrase, one can sample the scientific literature on any subject and get an unbiased view of the state of knowledge.

Figure 4.1 shows the results of an analysis of 928 abstracts, published in refereed journals during the period 1993 to 2003,

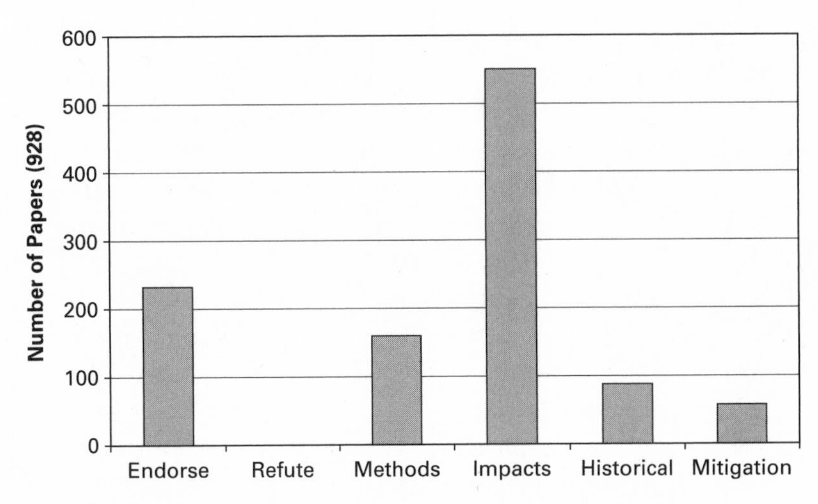

Figure 4.1
A Web of Science analysis of 928 abstracts using the keywords "global climate change." No papers in the sample provided scientific data to refute the consensus position on global climate change.

produced by a Web of Science search using the keyword phrase "global climate change."[5] After a first reading to determine appropriate categories of analysis, the papers were divided as follows: (1) those explicitly endorsing the consensus position, (2) those explicitly refuting the consensus position, (3) those discussing methods and techniques for measuring, monitoring, or predicting climate change, (4) those discussing potential or documenting actual impacts of climate change, (5) those dealing with paleoclimate change, and (6) those proposing mitigation strategies. How many fell into category 2—that is, how many of these papers present evidence that refutes the statement: "Global climate change is occurring, and human activities are at least part of the reason why"? The answer is remarkable: none.

A few comments are in order. First, often it is challenging to determine exactly what the authors of a paper do think about global climate change. This is a consequence of experts writing for experts: many elements are implicit. If a conclusion is widely accepted, then it is not necessary to reiterate it within the context of expert discussion. Scientists generally focus their discussions on questions that are still disputed or unanswered rather than on matters about which everyone agrees.

This is clearly the case with the largest portion of the papers examined (approximately half of the total)—those dealing with impacts of climate change. The authors evidently accept the premise that climate change is real and want to track, evaluate, and understand its impacts. Nevertheless, such impacts could, at least in some cases, be the results of natural variability rather than human activities. Strikingly, none of the papers used that possibility to argue against the consensus position.

Roughly 15 percent of the papers dealt with methods, and slightly less than 10 percent dealt with paleoclimate change. The most notable trend in the data is the recent increase in such papers; concerns about global climate change have given a boost to research in paleoclimatology and to the development of methods for measuring and evaluating global temperature and climate. Such papers are essentially neutral: developing better methods and understanding historic climate change are important tools for evaluating current effects, but they do not commit their authors to any particular opinion about those effects. Perhaps some of these authors are in fact skeptical of the current consensus, and this could be a motivation to work on a better understanding of the natural climate variability of the past. But again, none of the papers used that motivation to argue openly against the consensus, and it would be illogical if they did because a skeptical motivation does not constitute scientific evidence. Finally, approximately 20 percent of the papers explicitly endorsed the consensus position, and an additional 5 percent proposed mitigation strategies. In short, the basic reality of anthropogenic global climate change is no longer a subject of scientific debate.[6]

Some readers will be surprised by this result and wonder about the reliability of a study that failed to find any arguments against the consensus position when such arguments clearly exist. After all, anyone who watches the evening news or trolls the Internet knows that there is enormous debate about climate change, right? Well, no.

First, let's make clear what the scientific consensus is. It is over the reality of human-induced climate change. Scientists predicted a long time ago that increasing greenhouse gas emissions could change the climate, and now there is overwhelming evidence that it *is* changing the climate and that these changes are in addition to natural variability. Therefore, when contrarians try to shift the focus of attention to natural climate variability, they are misrepresenting the situation. No one denies the fact of natural variability, but natural variability alone does not explain what we are now experiencing. Scientists have also documented that some of the changes that are now occurring are clearly deleterious to both human communities and ecosystems (Arctic Council 2004). Because of global warming, humans are losing their homes and hunting grounds, and plants and animals are losing their habitats (e.g., Kolbert 2006; Flannery 2006).

Second, to say that global warming is real and happening now is not the same as agreeing about what will happen in the future. Much of the continuing debate in the scientific community involves the likely rate of future change. A good analogy is evolution. In the early twentieth century, paleontologist

George Gaylord Simpson introduced the concept of "tempo and mode" to describe questions about the manner of evolution—how fast and in what manner evolution proceeded. Biologists by the mid-twentieth century agreed about the reality of evolution, but there were extensive debates about its tempo and mode. So it is now with climate change. Virtually all professional climate scientists agree on the reality of human-induced climate change, but debate continues on tempo and mode.

Third, there is the question of what kind of dissent still exists. The analysis of the published literature presented here was done by sampling, using a keyword phrase that was intended to be fair, accurate, and neutral: "global climate change" (as opposed to, for example, "global warming," which might be viewed as biased). The *total* number of papers published over the last ten years having anything at all to do with climate change is probably over ten thousand, and no doubt some of the authors of the other over nine thousand papers have expressed skeptical or dissenting views. But the fact that the sample turned up no dissenting papers at all demonstrates that any remaining professional dissent is now exceedingly minor.

This suggests something discussed elsewhere in this book—that the mass media have paid a great deal of attention to a handful of dissenters in a manner that is greatly disproportionate with their representation in the scientific community. The number of climate scientists who actively do research in the field but disagree with the consensus position is evidently very small.

This is not to say that there are not a significant number of contrarians but to point out that most of them are not climate scientists and therefore have little (or no) basis to claim to be experts on the subjects on which they boldly pronounce. Some contrarians, like the physicist Frederick Seitz, were once active scientific researchers but have long since retired (and Seitz never actually did research in climate science; he was a solid-state physicist). Others, like the novelist Michael Crichton, are not scientists at all. What Seitz and Crichton have in common, along with most other contrarians, is that they do no new scientific research. They are not producing new evidence or new arguments. They are simply attacking the work of others and mostly doing so in the court of public opinion and in the mass media rather than in the halls of science.

This latter point is crucial and merits underscoring: the vast majority of materials denying the reality of global warming do not pass the most basic test for what it takes to be counted as scientific—namely, being published in a peer-reviewed journal. Contrarian views have been published in books and pamphlets issued by politically motivated think-tanks and widely spread across the Internet, but so have views promoting the reality of UFOs or the claim that Lee Harvey Oswald was an agent of the Soviet Union.

Moreover, some contrarian arguments are frankly disingenuous, giving the impression of refuting the scientific consensus when their own data do no such thing. One example will illustrate the point. In 2001, Willie Soon, a physicist at the Harvard-Smithsonian Center for Astrophysics, along with several colleagues, published a paper entitled "Modeling Climatic Effects of Anthropogenic Carbon Dioxide Emissions: Unknowns and Uncertainties" (Soon et al. 2001). This paper has been widely cited by contrarians as an important example of a legitimate dissenting scientific view published in a peer-review journal.[7] But the issue actually under discussion in the paper is how well models can predict the future—in other words, tempo and mode. The paper does not refute the consensus position, and the authors acknowledge this: "The purpose of [our] review of the deficiencies of climate model physics and the use of GCMs is to illuminate areas for improvement. Our review does not disprove a significant anthropogenic influence on global climate" (Soon et al. 2001, 259; see also Soon et al. 2002).

The authors needed to make this disclaimer because many contrarians do try to create the impression that arguments about tempo and mode undermine the whole picture of global climate change. But they don't. Indeed, one could reject all climate models and still accept the consensus position because models are only one part of the argument—one line of evidence among many.

Is there disagreement over the details of climate change? Yes. Are all the aspects of climate past and present well understood? No, but who has ever claimed that they were? Does climate science tell us what policy to pursue? Definitely not, but it does identify the problem, explain why it matters, and give society insights that can help to frame an efficacious policy response (e.g., Smith 2002).

So why does the public have the impression of disagreement among scientists? If the scientific community has forged a consensus, then why do so many Americans have the impression that there is serious scientific uncertainty about climate change?[8] There are several reasons. First, it is important to distinguish between scientific and political uncertainties. There are reasonable differences of opinion about how best to respond to climate change and even about how serious global warming is relative to other environmental and social issues. Some people have confused—or deliberately conflated—these two issues.

Scientists are in agreement about the reality of global climate change, but this does not tell us what to do about it.

Second, climate science involves prediction of future effects, which by definition is uncertain. It is important to distinguish among what is known to be happening now, what is likely to happen based on current scientific understanding, and what might happen in a worst-case scenario. This is not always easy to do, and scientists have not always been effective in making these distinctions. Uncertainties about the future are easily conflated with uncertainties about the current state of scientific knowledge.

Third, scientists have evidently not managed well enough to explain their arguments and evidence beyond their own expert communities. The scientific societies have tried to communicate to the public through their statements and reports on climate change, but what average citizen knows that the American Meteorological Society even exists or visits its home page to look for its climate-change statement?

There is also a deeper problem. Scientists are finely honed specialists trained to create new knowledge, but they have little training in how to communicate to broad audiences and even less in how to defend scientific work against determined and well-financed contrarians. Moreover, until recently, most scientists have not been particularly anxious to take the time to communicate their message broadly. Most scientists consider their "real" work to be the production of knowledge, not its dissemination, and often view these two activities as mutually exclusive. Some even sneer at colleagues who communicate to broader audiences, dismissing them as "popularizers."

If scientists do jump into the fray on a politically contested issue, they may be accused of "politicizing" the science and compromising their objectivity.[9] This places scientists in a double bind: the demands of objectivity suggest that they should keep aloof from contested issues, but if they don't get involved, no one will know what an objective view of the matter looks like. Scientists' reluctance to present their results to broad audiences has left scientific knowledge open to misrepresentation, and recent events show that there are plenty of people ready and willing to misrepresent it.

It's no secret that politically motivated think-tanks such as the American Enterprise Institute and the George Marshall Institute have been active for some time in trying to communicate a message that is at odds with the consensus scientific view (e.g., Gelbspan 1997, 2004). These organizations have successfully garnered a great deal of media attention for the tiny number of scientists who disagree with the mainstream view and for nonscientists, like novelist Michael Crichton, who pronounce loudly on scientific issues (Boykoff and Boykoff 2004).

This message of scientific uncertainty has been reinforced by the public relations campaigns of certain corporations with a large stake in the issue.[10] The most well known example is ExxonMobil, which in 2004 ran a highly visible advertising campaign on the op-ed page of the *New York Times*. Its carefully worded advertisements—written and formatted to look like newspaper columns and called op-ed pieces by ExxonMobil—suggested that climate science was far too uncertain to warrant action on it.[11] One advertisement concluded that the uncertainties and complexities of climate and weather means that "there is an ongoing need to support scientific research to inform decisions and guide policies" (Environmental Defense 2005). Not many would argue with this commonsense conclusion. But our scientists have concluded that existing research warrants that decisions and policies be made today.[12]

In any scientific debate, past or present, one can always find intellectual outliers who diverge from the consensus view. Even after plate tectonics was resoundingly accepted by earth scientists in the late 1960s, a handful of persistent resisters clung to the older views, and some idiosyncratics held to alternative theoretical positions, such as earth expansion. Some of these men were otherwise respected scientists, including Sir Harold Jeffreys, one of Britain's leading geophysicists, and Gordon J. F. MacDonald, a one-time science adviser to Presidents Lyndon Johnson and Richard Nixon; they both continued to reject plate tectonics until their dying day, which for MacDonald was in 2002. Does that mean that scientists should reject plate tectonics, that disaster-preparedness campaigns should not use plate-tectonics theory to estimate regional earthquake risk, or that schoolteachers should give equal time in science classrooms to the theory of earth expansion? Of course not. That would be silly and a waste of time.

No scientific conclusion can ever be proven, and new evidence may lead scientists to change their views, but it is no more a "belief" to say that earth is heating up than to say that continents move, that germs cause disease, that DNA carries hereditary information, and that HIV causes AIDS. You can always find someone, somewhere, to disagree, but these conclusions represent our best current understandings and therefore our best basis for reasoned action (Oreskes 2004).

How Do We Know We're Not Wrong?

Might the consensus on climate change be wrong? Yes, it could be, and if scientific research continues, it is almost certain that some aspects of the current understanding will be modified, perhaps in significant ways. This possibility can't be denied.

The relevant question for us as citizens is not whether this scientific consensus *might* be mistaken but rather whether there is any reason to think that it *is* mistaken.

How can outsiders evaluate the robustness of any particular body of scientific knowledge? Many people expect a simple answer to this question. Perhaps they were taught in school that scientists follow "the scientific method" to get correct answers, and they have heard some climate-change deniers suggesting that climate scientists do not follow the scientific method (because they rely on models, rather than laboratory experiments) so their results are suspect. These views are wrong.

Contrary to popular opinion, there is no scientific method (singular). Despite heroic efforts by historians, philosophers, and sociologists, there is no answer to what the methods and standards of science really are (or even what they should be). There is no methodological litmus test for scientific reliability and no single method that guarantees valid conclusions that will stand up to all future scrutiny.

A positive way of saying this is that scientists have used a variety of methods and standards to good effect and that philosophers have proposed various helpful criteria for evaluating the methods used by scientists. None is a magic bullet, but each can be useful for thinking about what makes scientific information a reliable basis for action.[13] How does current scientific knowledge about climate stand up to these diverse models of scientific reliability?

The Inductive and Deductive Models of Science

The most widely cited models for understanding scientific reasoning are induction and deduction. *Induction* is the process of generalizing from specific examples. If I see 100 swans and they are all white, I might conclude that all swans are white. If

I saw 1,000 white swans or 10,000, I would surely think that all swans were white, yet a black one might still be lurking somewhere. As David Hume famously put it, even though the sun has risen thousands of times before, we have no way to prove that it will rise again tomorrow.

Nevertheless, common sense tells us that the sun is extremely likely to rise again tomorrow, even if we can't logically prove that it's so. Common sense similarly tells us that if we had seen ten thousand white swans, then our conclusion that all swans were white would be more robust than if we had seen only ten. Other things being equal, the more we know about a subject, and the longer we have studied it, the more likely our conclusions about it are to be true.

How does climate science stand up to the inductive model? Does climate science rest on a strong inductive base? Yes. Humans have been making temperature records consistently for over 150 years, and nearly all scientists who have looked carefully at these records see an overall increase since the industrial revolution about 0.6° to 0.7°C (1.1° to 1.3°F) (Houghton, Jenkins, and Ephraums 1990; Bruce et al. 1996; Watson et al. 1996; McCarthy et al. 2001; Houghton et al. 2001; Metz et al. 2001; Watson 2001; Weart 2003). The empirical signal is clear, even if not all the details are clear.

How reliable are the early records? How do you average the data to be representative of the globe as a whole, even though much of the early data comes from only a few places, mostly in Europe? Scientists have spent quite a bit of time addressing these questions; most have satisfied themselves that the empirical signal is clear. But even if scientists doubted the older records, the more recent data show a strong increase in temperatures over the past thirty to forty years, just when the amount of carbon dioxide and other greenhouses gases in the atmosphere was growing dramatically (McCarthy et al. 2001; Houghton 2001; Metz et al. 2001; Watson 2001).

Moreover these records—based on measurements with instruments, such as thermometers—are corroborated by independent evidence from tree rings, ice cores, and coral reefs. A recent paper by Jan Esper at the Swiss Federal Research Center and colleagues at Columbia University, shows, for example, that tree rings can provide a reliable, long-term record of temperature variability that largely agrees with the instrumental records over the past 150 years (Esper, Cook, and Schweingruber 2003).

While many scientists are happy simply to obtain consistent results—often no trivial task—others may deem it important to find some means to test whether their conclusions are right. This has led to the view that the core of scientific method is testing theories through logical deductions.

Deduction is drawing logical inferences from a set of premises—the stock-in-trade of Sherlock Holmes. In science, deduction is generally presumed to work as part of what has come to be known as the *hypothetico-deductive model*—the model you will find in most textbooks that claim to teach the scientific method. In this view, scientists develop hypotheses and then test them. Every hypothesis has logical consequences—deductions—and one can try to determine whether the deductions are correct. If they are, they support the hypothesis. If they are not, then the hypothesis must be revised or rejected. It's especially good if the prediction is something that would otherwise be quite unexpected because that would suggest that it didn't just happen by chance.

The most famous example of successful deduction in the history of science is the case of Ignaz Semmelweis, who in the 1840s deduced the importance of hand washing to prevent the spread of infection (Gillispie 1975; Hempel 1965). Semmelweis had noticed that many women were dying of fever after giving birth at his Viennese hospital. Surprisingly, women who had their infants on the way to the hospital—seemingly under more adverse conditions—rarely died of fever. Nor did women who gave birth at another hospital clinic where they were attended by midwives. Semmelweis was deeply troubled by this.

In 1847, a friend of Semmelweis, Jakob Kolletschka, cut his finger while doing an autopsy and soon died. Autopsy revealed a pathology very similar to the women who had died after childbirth; something in the cadaver had apparently caused his death. Semmelweis knew that many of the doctors at his clinic routinely went directly from conducting autopsies to attending births, but midwives did not perform autopsies, so he hypothesized that the doctors were carrying cadaveric material on their hands, which was infecting the women (and killed his friend). He deduced that if physicians washed their hands before attending the women, then the infection rate would decline. They did so, and the infection rate did decline, demonstrating the power of the hypothetico-deductive method.

How does climate science stand up to this standard? Have climate scientists made predictions that have come true? Absolutely. The most obvious is the fact of global warming itself. As already has been noted in previous chapters, scientific concern over the effects of increased atmospheric carbon dioxide is based on physics—the fact that CO_2 is a greenhouse gas. In the early twentieth century, Swedish chemist Svante Arrhenius predicted that increasing carbon dioxide from the burning of fossil fuels would lead to global warming, and by midcentury, a number of other scientists, including G. S. Callendar, Roger Revelle, and Hans Suess, concluded that the effect might soon be quite noticeable, leading to sea level rise and other global changes. In 1965, Revelle and his colleagues wrote, "By the year 2000, the increase in atmospheric CO_2 . . . may be sufficient to produce measurable and perhaps marked change in climate, and will almost certainly cause significant changes in the temperature and other properties of the stratosphere" (Revelle 1965, 9). This prediction has come true (Fleming 1998; Weart 2003; McCarthy et al. 2001; Houghton et al. 2001; Metz et al. 2001; Watson 2001).

Another prediction fits the category of something unusual that you might not even think of without the relevant theory. In 1980, climatologist Suki Manabe predicted that the effects of global warming would be strongest first in the polar regions. *Polar amplification* was not an induction from observations but a deduction from theoretical principles: the notion of ice-albedo feedback. The reflectivity of a material is called its *albedo*. Ice has a high albedo. It reflects sunlight back into space much more effectively than grass, dirt, or water, and one reason polar regions are as cold as they are is that snow and ice are very effective in reflecting solar radiation back into space. But if the snow starts to melt and bare ground (or water) is exposed, the reflection effect diminishes. Less ice means less reflection, which means more solar heat is absorbed, leading to yet more melting in a positive feedback loop. So once warming begins, its effects are more pronounced in polar regions than in temperate ones. The Arctic Climate Impact Assessment concluded in 2004 that this prediction has also come true (Manabe and Stouffer 1980, 1994; Holland and Bitz 2003; Arctic Council 2004).

Falsificationism

Ignaz Semmelweis is among the famous figures in the history of science because his work in the 1840s foreshadows the germ theory of disease and the saving of millions of human lives. But the story has a twist because Semmelweis was right for the wrong reason. Cadaveric matter was *not* the cause of the infections: germs were. In later years, this would be demonstrated by James Lister, Robert Koch, and Louis Pasteur, who realized that hand washing was effective not because it removed the cadaveric material but because it removed the germs associated with that material.

The story illustrates the fundamental logical flaw with the hypothetico-deductive model—the fallacy of affirming the consequent. If I make a prediction, and it comes true, it does not prove that my hypothesis was correct; my prediction may have come true for other reasons. The other reasons may be related to the hypothesis—germs *were* associated with cadaveric matter—but in other cases the connection may be entirely coincidental. I can convince myself that I have proved my theory right, but this would be self-deception. This realization led the twentieth-century philosopher Karl Popper to suggest that you can never prove a theory true but you can prove it false—a view known as *falsificationism* (Popper 1959).

How does climate science hold up to this modification? Can climate models be refuted? Falsification is a bit of a problem for all models—not just climate models—because many models are built to forecast the future and the results will not be known for some time. By the time we find out whether the long-term predictions of a model are right or wrong, that knowledge won't be of much use. For this reason, many models are tested by seeing if they can accurately reproduce past events. In principle, this should be an excellent test—a climate model that failed to reproduce past temperature records might be considered falsified—but in reality, it doesn't work quite that way.

Climate models are complex, and they involve many variables—some that are well measured and others that are not. If a model does not reproduce past data very well, most modelers assume that one or more of the model parameters are not quite right, and they make adjustments in an attempt to obtain a better fit. This is generally referred to as *model calibration*, and many modelers consider it an essential part of the process of building a good model. But the problem is that calibration can make models refutation-proof: the model doesn't get rejected; it gets revised. If model results were the only basis for current scientific understanding, they would be grounds for some healthy skepticism. Models are therefore best viewed as heuristic devices: a means to explore what-if scenarios. This is, indeed, how most modelers use them: to answer questions like "If we double the amount of CO_2 in the atmosphere, what is the most likely outcome?"

One way in which modelers address the fact that a model can't be proved right or wrong is to make lots of different models that explore diverse possible outcomes—what modelers call *ensembles*. An example of this is ⟨climateprediction.net⟩, a Web-based mass-participation experiment that enlists members of the public to run climate models on their home computers to explore the range of likely and possible climate outcomes under a variety of plausible conditions.

Over ninety thousand participants from over 140 countries have produced tens of thousands of runs of a general circulation model produced by the Hadley Centre for Climate Prediction and Research. Figure 4.2 presents some initial results, published in the journal *Nature* in 2005, for a steady-state model in which atmospheric carbon dioxide is doubled relative to preindustrial levels and the model earth is allowed to adjust.

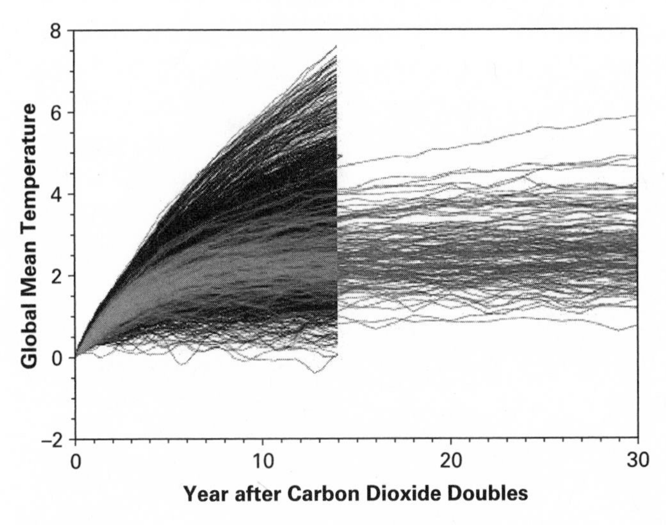

Figure 4.2
Changes in global mean surface temperature (C°) after carbon dioxide values in the atmosphere are doubled. The black lines show the results of 2,579 fifteen-year simulations by members of the general public using their own personal computers. The grey lines show comparable results from 127 thirty-year simulations completed by Hadley Centre scientists on the Met Office's supercomputer (⟨www.metoffive.gov.uk⟩). Figure prepared by Ben Sanderson with help from the ⟨climateprediction.net⟩ project team.
Source: Reproduced by permission from ⟨http://www.climateprediction.net/science/results_cop10.php⟩.

The results in black are the climateprediction.net's mass-participation runs; the results in grey come from runs made by professional climate scientists at the Hadley Centre on a supercomputer (Stainforth et al. 2005).

What does an ensemble like this show? For one thing, no matter how many times you run the model, you almost always get the same qualitative result: the earth will warm. The unanswered question is how much and how fast—in other words, tempo and mode.

The models vary quite a bit in their tempo and mode, but nearly all fall within a temperature range of 2° to 8°C (4° to 14°F) within fifteen years after the earth's atmosphere reaches a doubling of atmospheric CO_2. Moreover, most of the runs are still warming at that point. The model runs were stopped at year 15 for practicality, but most of them had not yet reached equilibrium: model temperatures were still rising. Look again at figure 4.2. If the general-public model runs had been allowed to continue out to thirty years, as the Hadley Centre scientists' model runs do, many of them would apparently have reached still higher temperatures, perhaps as high as 12°C.

How soon will our atmosphere reach a CO_2 level of twice the preindustrial level? The answer depends largely on how much carbon dioxide we humans put into the atmosphere—a parameter that cannot be predicted by a climate model. Note also that in these models CO_2 does not continue to rise: it is fixed at twice preindustrial levels. Most experts believe that unless major steps are taken quickly, atmospheric CO_2 levels will go well above that level. If CO_2 triples or quadruples, then the expected temperature increase will also increase. No one can say precisely when earth's temperature will increase by any specific value, but the models indicate that it almost surely *will* increase. With very few exceptions, the models show the earth warming, and some of them show the earth warming very quickly.

Is it possible that *all* these model runs are wrong? Yes, because they are variations on a theme. If the basic model conceptualization was wrong in some way, then all the models runs would be wrong. Perhaps there is a negative feedback loop that we have not yet recognized. Perhaps the oceans can absorb more CO_2 than we think, or we have missed some other carbon sink (Smith 2002). This is one reason that continued scientific investigation is warranted. But note that Svante Arrhenius and Guy Callendar predicted global warming before anyone ever built a global circulation model (or even had a digital computer). Climate models give us a tool for exploring scenarios and interactions, but you don't need a climate model to know that global warming is a real problem.

If climate science stands with or without climate models, then is there any information that would show that climate science is wrong? Sure. Scientists might discover a mistake in their basic physical understanding that showed they had misconceptualized the whole issue. They could discover that they had overestimated the significance of carbon dioxide and underestimated the significance of some other parameter. But if such mistakes are found, there is no guarantee that correcting them will lead to a more optimistic scenario. It could well be the case that scientists discover neglected factors that show that the problem is even worse than we'd supposed.

Moreover, there is another way to think about this issue. Contrarians have put inordinate amounts of effort into trying to find something that is wrong with climate science, and despite all this effort, they have come up empty-handed. Year after year, the evidence that global warming is real and serious has only strengthened.[14] Perhaps that is the strongest argument of all. Contrarians have repeatedly tried to falsify the consensus, and they have repeatedly failed.

Consilience of Evidence

Most philosophers and historians of science agree that there is no iron-clad means to prove a scientific theory. But if science does not provide proof, then what is the purpose of induction, hypothesis testing, and falsification? Most would answer that, in various ways, these activities provide warrant for our views. Do they?

An older view, which has come back into fashion of late, is that scientists look for consilience of evidence. *Consilience* means "coming together," and its use is generally credited to the English philosopher William Whewell, who defined it as the process by which sets of data—independently derived—coincided and came to be understood as explicable by the same theoretical account (Gillispie 1981; Wilson 1998). The idea is not so different from what happens in a legal case. To prove a defendant guilty beyond a reasonable doubt, a prosecutor must present a variety of evidence that holds together in a consistent story. The defense, in contrast, might need to show only that some element of the story is at odds with another to sow reasonable doubt in the minds of the jurors. In other words, scientists are more like lawyers than they might like to admit. They look for independent lines of evidence that hold together.

Do climate scientists have a consilience of evidence? Again the answer is yes. Instrumental records, tree rings, ice cores, borehole data, and coral reefs all point to the same conclusion: things are getting warmer overall. Keith Briffa and Timothy Osborn of the Climate Research Unit of the University of East Anglia compared Esper's tree-ring analysis with six other reconstructions of global temperature between the years 1000 and 2000 (Briffa and Osborn 2002). All seven analyses agree: temperatures increased dramatically in the late twentieth century relative to the entire record of the previous millennium. Temperatures vary naturally, of course, but the absolute magnitude of global temperatures in the late twentieth century was higher than *any* known temperatures in the previous one thousand years, and many different lines of evidence point in this direction.

Inference to the Best Explanation

The various problems in trying to develop an account of how and why scientific knowledge is reliable have led some philosophers to conclude that the purpose of science is not proof, but explanation. Not just any explanation will do, however; the best explanation is the one that is consistent with the evidence (e.g., Lipton 1991). Certainly, it is possible that a malicious or mischievous deity placed fossils throughout the geological record to trick us into believing organic evolution, but to a scientist this is not the best explanation because it invokes supernatural effects, and the supernatural is beyond the scope of scientific explanation. (It might not be the best explanation to a theologian, either, if that theologian was committed to heavenly benevolence.) Similarly, I might try to explain the drift of the continents through the theory of the expanding earth—as some scientists did in the 1950s—but this would not be the best explanation because it fails to explain why the earth has conspicuous zones of compression as well as tension. The philosopher of science Peter Lipton has put it this way: every set of facts has a diversity of possible explanations, but "we cannot infer something simply because it is a possible explanation. It must somehow be the best of competing explanations" (Lipton 2004, 56).

Best is a term of judgment, so it doesn't entirely solve our problem, but it gets us thinking about what it means for a scientific explanation to be the best available—or even just a good one. It also invites us to ask the question, "Best for what purpose?" For philosophers, *best* generally means that an explanation is consistent with all the available evidence (not just selected portions of it), that the explanation is consistent with other known laws of nature and other bodies of accepted evidence (and not in conflict with them), and that the explanation does not invoke supernatural events or causes that virtually by definition cannot be refuted. In other words, *best* can be judged in terms of the various criterion invoked by all the models of science discussed above: Is there an inductive basis? Does the theory pass deductive tests? Do the various elements of the theory fit with each other and with other established scientific information? And is the explanation *scientific* in the sense of being potentially refutable and not invoking unknown, inexplicable, or supernatural causes?

Contrarians have tried to suggest that the climate effects we are experiencing are simply natural variability. Climate does vary, so this is a possible explanation. No one denies that. But is it the *best* explanation for what is happening now? Most climate scientists would say that it's not the best explanation. In fact, it's not even a good explanation—because it is inconsistent with much of what we know.

Should we believe that the global increase in atmospheric carbon dioxide has had a negligible effect even though basic physics indicates otherwise? Should we believe that the correlation between increased CO_2 and increased temperature is just a weird coincidence? If there were no theoretical reason to relate them and if Arrhenius, Callendar, Suess, and Revelle had not predicted that all this would all happen, then one might well conclude that rising CO_2 and rising temperature were merely coincidental. But we have every reason to believe that there is a causal connection and no good reason to believe that it is a coincidence. Indeed, the only reason we might think otherwise is to avoid committing to action: if this is just a natural cycle in which humans have played no role, then maybe global warming will go away on its own in due course.

And that sums up the problem. To deny that global warming is real is precisely to deny that humans have become geological agents, changing the most basic physical processes of the earth. For centuries, scientists thought that earth processes were so large and powerful that nothing we could do would change them. This was a basic tenet of geological science: that human chronologies were insignificant compared with the vastness of geological time; that human activities were insignificant compared with the force of geological processes. And once they were. But no more. There are now so many of us cutting down so many trees and burning so many billions of tons of fossil fuels that we have indeed become geological agents. We have changed the chemistry of our atmosphere, causing sea level to rise, ice to melt, and climate to change. There is no reason to think otherwise.

Notes

1. Contrast this with the results of the Intergovernmental Panel on Climate Change's *Third* and *Fourth Assessment Reports*, which state unequivocally that average global temperatures have risen (Houghton et al. 2001; Alley et al. 2007).

2. It must be acknowledged that in any area of human endeavor, leadership may diverge from the views of the led. For example, many Catholic priests endorse the idea that priests should be permitted to marry (Watkin 2004).

3. In recent years, climate-change deniers have increasingly turned to nonscientific literature as a way to promulgate views that are rejected by most scientists (see, for example, Deming 2005).

4. An e-mail inquiry to the Thomson Scientific Customer Technical Help Desk produced this reply: "We index the following number of

papers in Science Citation Index—2004, 1,057,061 papers; 2003, 1,111,398 papers."

5. The analysis begins in 1993 because that is the first year for which the database consistently published abstracts. Some abstracts initially compiled were deleted from our analysis because the authors of those papers had put "global climate change" in their key words, but their papers were not actually on the subject.

6. This is consistent with the analysis of historian Spencer Weart, who concluded that scientists achieved consensus in 1995 (see Weart 2003).

7. In e-mails that I received after publishing my essay in *Science* (Oreskes 2004), this paper was frequently invoked.

8. And we do. According to *Time* magazine, a recent Gallup poll reported that "64 percent of Americans think scientists disagree with one another about global warming" (Americans see a climate problem 2006).

9. Objectivity certainly can be compromised when scientists address charged issues. This is not an abstract concern. It has been demonstrated that scientists who accept research funds from the tobacco industry are much more likely to publish research results that deny or downplay the hazards of smoking than those who get their funds from the National Institutes of Health, the American Cancer Society, or other nonprofit agencies (Bero 2003). On the other hand, there is a large difference between accepting funds from a patron with a clearly vested interest in a particular epistemic outcome and simply trying one's best to communicate the results of one's research clearly and in plain English.

10. Some petroleum companies, such as BP and Shell, have refrained from participating in misinformation campaigns (see Browne 1997). Browne began his 1997 lecture by focusing on what he accepted as "two stark facts. The concentration of carbon dioxide in the atmosphere is rising, and the temperature of the Earth's surface is increasing." For an analysis of diverse corporate responses, see Van den Hove et al. (2003).

11. For an analysis of one ad, "Weather and Climate," see Environmental Defense (2005). An interesting development in 2003 was that Institutional Shareholders Services advised ExxonMobil shareholders to ask the company to explain its stance on climate-change issues and

to divulge financial risks that could be associated with it (see ISS in favor of ExxonMobil 2003).

12. These efforts to generate an aura of uncertainty and disagreement have had an effect. This issue has been studied in detail by academic researchers (see, for example, Boykoff and Boykoff 2004).

13. *Reliable* is a term of judgment. By *reliable basis for action*, I mean that it will not lead us far astray in pursuing our goals, or if it does lead us astray, at least we will be able to look back and say honestly that we did the best we could given what we knew at the time.

14. This is evident when the three IPCC assessments—1990, 1995, 2001—are compared (Houghton et al. 1990; Bruce et al. 1996; Watson et al. 1996; Houghton et al. 2001; Metz et al. 2001; Watson 2001; see also Weart 2003).

References

Alley, Richard et al. 2007. *Climate change 2007: The physical science basis: Summary for policymakers*. Intergovernmental Panel on Climate Change. Geneva: IPCC Secretariat. ⟨http://www.ipcc.ch/SPM2feb07.pdf⟩ (accessed March 31, 2007).

American Geophysical Union Council. 2003. Human impacts of climate. American Geophysical Union, Washington, DC. ⟨www.agu.org/sci_soc/policy/climate_change_position.html⟩ (accessed March 1, 2005).

American Meteorological Society. 2003. Climate change research: Issues for the atmospheric and related sciences. *Bulletin of the American Meteorological Society* 84. ⟨http://www.ametsoc.org/policy/climatechangeresearch_2003.html⟩ (accessed May 1, 2006).

Americans see a climate problem. *Time.com*, March 26. ⟨http://www.time.com/time/nation/article/0,8599,1176967,00.html⟩ (accessed May 1, 2006).

Arctic Council. 2004. *Arctic climate impact assessment*. Arctic Council, Oslo, Norway. ⟨http://www.acia.uaf.edu⟩ (accessed March 14, 2005).

Bero L. 2003. Implications of the tobacco industry documents for public health and policy. *Annual Review of Public Health* 24: 267–88.

Boykoff, M. T., and J. M. Boykoff. 2004. Balance as bias: Global warming and the U.S. prestige press. *Global Environmental Change* 14: 125–36.

Briffa, K. R., and T. J. Osborn. 2002. Blowing hot and cold. *Science* 295: 2227–28.

Browne, E. J. P. 1997. Climate change: The new agenda. Paper presented at Stanford University, May 19, Group Media and Publications, British Petroleum Company.

Bruce, James P., Hoesung Lee, and Erik F. Haites, eds. 1996. *Climate change 1995: Economic and social dimensions of climate change.* Intergovernmental Panel on Climate Change. Cambridge: Cambridge University Press.

Deming, David. 2005. How "consensus" on global warming is used to justify draconian reform. *Investor's Business Daily*, March 18, A16.

Environmental Defense. 2005. Too slick: Stop ExxonMobil's global warming misinformation campaign. ⟨http://actionnetwork.org/campaign/exxonmobil?source=edac2⟩ (accessed March 14, 2005).

Esper, J., E. R. Cook, and F. H. Schweingruber. 2003. Low-frequency signals in long tree-ring chronologies for reconstructing past temperature variability. *Science* 295: 2250–53.

Flannery, Tim. 2006. *The weather makers: How man is changing the climate and what it means for life on earth.* New York: Atlantic Monthly Press.

Fleming, James Rodger. 1998. *Historical Perspectives on Climate Change.* New York: Oxford University Press.

Gelbspan, Ross. 1997. *The heat is on: The high stakes battle over earth's threatened climate.* Reading, MA: Addison-Wesley.

Gelbspan, Ross. 2004. *Boiling point: How politicians, big oil and coal, journalists, and activists are fueling the climate crisis—and what we can do to avert disaster.* New York: Basic Books.

Gillispie, Charles C., ed. 1975. Semmelweis. In *Dictionary of scientific biography.* Vol. 12. New York: Scribner.

Gillispie, Charles C., ed. 1981. *Dictionary of scientific biography.* Vol. 12. New York: Scribner.

Harrison, Paul, and Fred Pearce. 2000. *AAAS atlas of population and environment.* Berkeley: University of California Press. ⟨http://www.ourplanet.com/aaas/pages/atmos02.html⟩ (accessed May 1, 2006).

Hempel, Carl. 1965. *Aspects of scientific explanation, and other essays in the philosophy of science*. New York: Free Press.

Holland, M. M., and C. M. Bitz. 2003. Polar amplification of climate change in coupled models. *Climate Dynamics* 21: 221–32.

Houghton, J. T., G. J. Jenkins, and J. J. Ephraums, eds. 1990. *Scientific assessment of climate change: Report of Working Group I*. Intergovernmental Panel on Climate Change. Cambridge: Cambridge University Press.

Houghton, J. T., Y. Ding, D. J. Griggs, M. Noguer, P. J. van der Linden, X. Dai, K. Maskell, and C. A. Johnson, eds. 2001. *Climate change 2001: The scientific basis* (Third Assessment Report). Intergovernmental Panel on Climate Change. Cambridge: Cambridge University Press.

Intergovernmental Panel on Climate Change (IPCC). 2005. About IPCC, February 7. ⟨http://www.ipcc.ch/about/about.htm⟩ (accessed March 1, 2005).

ISS in favor of ExxonMobil global warming proposals. *Planet Ark*, May 19. ⟨http://www.planetark.com/dailynewsstory.cfm/newsid/20824/story.htm⟩ (accessed March 14, 2005).

Kolbert, Elizabeth. 2006. *Field notes from a catastrophe*. New York: Bloomsbury.

Lipton, Peter. 1991. *Inference to the best explanation*. Oxford: Routledge.

Lipton, Peter. 2004. *Inference to the best explanation*. 2nd ed. Oxford: Routledge.

Manabe, S., and R. J. Stouffer. 1980. Sensitivity of a global climate model to an increase of CO_2 concentration in the atmosphere. *Journal of Geophysical Research* 85(C10): 5529–54.

Manabe, S., and R. J. Stouffer. 1994. Multiple-century response of a coupled ocean-atmosphere model to an increase of atmospheric carbon dioxide. *Journal of Climate* 7(1): 5–23.

McCarthy, James J., Osvaldo F. Canziani, Neil A. Leary, David J. Dokken, and Kasey S. White, eds. 2001. *Climate change 2001: Impacts, adaptation and vulnerability*. Intergovernmental Panel on Climate Change. Cambridge: Cambridge University Press.

Metz, Bert, Ogunlade Davidson, Rob Swart, and Jiahua Pan, eds. 2001. *Climate change 2001: Mitigation*. Intergovernmental Panel on Climate Change. Cambridge: Cambridge University Press.

National Academy of Sciences, Committee on the Science of Climate Change. 2001. *Climate change science: An analysis of some key questions*. Washington, DC: National Academy Press.

Oreskes, N. 2004. Beyond the ivory tower: The scientific consensus on climate change. *Science* 306(5702): 1686.

Popper, K. R. 1959. *The logic of scientific discovery*. London: Hutchinson.

Price, Derek de Solla. 1986. *Little science, big science—and beyond*. New York: Columbia University Press.

Revelle, Roger. 1965. Atmospheric carbon dioxide. In *Restoring the quality of our environment: A report of the Environmental Pollution Panel*, 111–33. Washington, DC: President's Science Advisory Committee.

Roach, John. 2004. The year global warming got respect. *National Geographic*, December 29. ⟨http://news.nationalgeographic.com/news/2004/12/1229_041229_climate_change_consensus.html⟩ (accessed February 24, 2005).

Smith, Leonard A. 2002. What might we learn from climate forecasts? *Proceedings of the National Academy of Sciences* 99 (suppl. 1): 2487–92.

Soon, W., S. Baliunas, S. B. Idso, K. Y. Kondratyev, and E. S. Posmentier. 2001. Modeling climatic effects of anthropogenic carbon dioxide emissions: Unknowns and uncertainties. *Climate Research* 18: 259–75.

Soon, W., S. Baliunas, S. B. Idso, K. Y. Kondratyev, and E. S. Posmentier. 2002. Modeling climatic effects of anthropogenic carbon dioxide emissions: Unknowns and uncertainties, reply to Risbey. *Climate Research* 22: 187–88.

Stainforth, D., T. Aina, C. Christensen, M. Collins, N. Faull, D. J. Frame, J. A. Kettleborough, S. Knight, A. Martin, J. M. Murphy, C. Piani, D. Sexton, L. A. Smith, R. A. Spicer, A. J. Thorpe, and M. R. Allen. 2005. Uncertainty in predictions of the climate response to rising levels of greenhouse gases. *Nature* 433: 403–06.

Van den Hove, Sybille, Marc Le Menestrel, and Henri-Claude de Bettignies. 2003. The oil industry and climate change: Strategies and ethical dilemmas. *Climate Policy* 2: 3–18.

Watkin, Daniel J. 2004. Roman Catholic priests' group calls for allowing married clergy members. *New York Times*, April 28, B5.

Watson, Robert T., ed. 2001. *Climate change 2001: Synthesis report.* Intergovernmental Panel on Climate Change. Cambridge: Cambridge University Press.

Watson, R. T., Marufu C. Zinyowera, and Richard H. Moss, eds. 1996. *Climate change 1995: Impacts, adaptations and mitigation of climate change—Scientific-technical analyses.* Intergovernmental Panel on Climate Change. Cambridge: Cambridge University Press.

Weart, Spencer R. 2003. *The discovery of global warming.* Cambridge, MA: Harvard University Press.

Wilson, Edward O. 1998. *Consilience: The unity of knowledge.* New York: Alfred A. Knopf.

5

Climate Change: How the World Is Responding

Joseph F. C. DiMento and Pamela Doughman

Scientists warn that it is too late to avoid climate change, but we can act now to ease the problem for our children and grandchildren. This realization has been made by a large part of the world. A 2006 *Los Angeles Times*/Bloomberg survey found that almost half of Americans think global warming is caused more by human activities than by natural changes in the climate and that 56 percent believe that the government could do more to address the problem (Boxall 2006). They are basing their opinions on the evidence presented by an army of climate scientists and an almost equally large group of policy analysts, politicians, and other government leaders.

Recognition that meaningful actions to protect our climate are possible has come in different forms across the globe—modest binding commitments by the developed world to reduce greenhouse-gas emissions, voluntary efforts, and efforts limited to the major sources of emissions. Most countries have concluded that meeting regularly to review current science, policies, and programs to decrease greenhouse-gas emissions makes sense. Perhaps meetings are not the most efficient way to address a global environmental challenge, but they are one of the few options that countries have to face this type of problem.

In their official positions, countries have moved with varying senses of urgency. Some are fearful that their land mass and resource base are threatened in the near future. Others have concluded that economic growth and other environmental challenges are more important than climate change. Several nations have gone beyond the official international requirements and have taken steps to make significant reductions in their contributions to climate change. The link between climate policy and other national policies (such as energy independence, technological leadership, and other green goals) has been recognized, so climate-change policy might achieve several important national objectives. States and provinces have recognized the benefits of some climate-change policies, as have utilities, high-tech companies, and even some energy giants.

These conclusions have come in fits and starts with significant shifts in the positions of important countries, including the United States, and disagreements within countries and within states and businesses about how many changes they should make. The costs of change are important, and internationally they have been calculated in different ways and balanced differently against the benefits of reducing emissions of greenhouse gases and against the benefits of moving slowly. The understanding that major shifts in climate-change policy will produce winners and losers has set the political stage for world opinion.

Independent of official international and national responses, smaller organizations and even individuals have turned into actions their beliefs that climate change can be mitigated. Those actions reflect very different understandings of how important global warming is compared to other social goals. In the 2006 national survey, for example, even those who favored further government action did not favor capping emissions from vehicles and businesses: only 11 percent favored those ideas (Boxall 2006).

In this chapter, we describe how the world has reacted to the scientific message. We focus on the supporters and opponents of international efforts to control greenhouse emissions through international laws—why there have been such different understandings of what (if anything) needs to be done, what U.S. climate-change policy has been, and what the responses of state and local governments and industry have been.

Efforts at International Cooperation and a Legal Response

In the face of mounting evidence that action was needed to reduce the growth of greenhouse gases in the atmosphere, ten thousand delegates from around the world met in Kyoto, Japan, in 1997 to make good on commitments made by the United Nations Framework Convention on Climate Change (FCCC) in 1992. After a series of late-night negotiations, the delegates produced an international agreement—the Kyoto Protocol—that called on industrialized countries to reduce emissions of greenhouse gases in 2008 to 2012 to at least 5 percent below 1990 levels.

Some thought the Kyoto Protocol was a promising start to a difficult problem; others wavered. The United States thought it was a promising start, then wavered, then reneged. Russia hesitated but then supported the Kyoto Protocol, which allowed the agreement to go into effect in 2005. In twenty-five years of international efforts to address climate change, the 1997 Kyoto Protocol marked the first agreement on binding limits of greenhouse-gas emissions.

Pre-Kyoto

Before the 1997 Kyoto Protocol, the dominant focus of international environmental policy was on cooperative research and voluntary goals for reducing emissions. In 1979, the First World Climate Conference called for greater cooperation to study climate change and prevent it from getting worse (IUCC 1979).

The late 1980s were a period of growing awareness of global ecological problems and their social, economic, and health dilemmas. Scientists discovered the weakening of the earth's protective ozone layer. An international agreement, the Montreal Protocol, was signed in 1987 to control emissions of ozone-depleting substances. The publication of *Our Common Future* by the World Commission on Environment and Development (Bruntland 1987) focused international attention on the need for sustainable development and on the links among the environment, society, and the economy. The book emphasized that damaging any one of these three elements weakens the other two—either in the current generation or in those that follow. Awareness of climate change was also heightened during this time period, through publication of the First Assessment Report of the Intergovernmental Panel on Climate Change (IPCC 1990). This report said that human activities are substantially increasing the amount of greenhouse gases in the atmosphere. Although effects vary according to region and time period, the report predicted that the average temperature of the earth's surface will become warmer than the average earth surface temperatures known during the past ten thousand years.

This period's sense of urgency was reflected in the frequency and magnitude of the international environmental negotiations on climate change that were conducted during this time (see table 5.1). Although eleven years passed between the First World Climate Conference in 1979 and the Second World Climate Conference in 1990, a flurry of negotiations led to the 1992 United Nations Conference on Environment and Development in Rio de Janeiro, Brazil. Several international environmental agreements were signed at the Rio Conference, including the Framework Convention on Climate Change. The FCCC set out general parameters and principles for international efforts to address climate change; details were to be worked out in subsequent negotiations. The FCCC aims for "stabilization of greenhouse gas concentrations in the atmosphere at a level that would prevent dangerous anthropogenic interference with the climate system. This level should be achieved within a time-frame sufficient to allow ecosystems to adapt naturally to climate change, to ensure that food production is not threatened and to enable economic development to proceed in a sustainable manner."

In support of the idea that climate change is a common concern and the world's nations have different responsibilities to the earth's environment, developed countries and developing countries have different commitments under the FCCC. All signers agreed to provide information and work together, but developed countries agreed to be leaders in addressing climate change. They decided to reduce human emissions of carbon dioxide and other greenhouse gases not controlled by the Montreal Protocol to 1990 levels by 2000 through the use of appropriate and cost-effective technologies and means. Reductions can be achieved individually or jointly. The FCCC encourages developing countries to reduce greenhouse-gas emissions, and to promote sustainable development in developing countries, developed countries agreed to fund developing countries' compliance with the treaty.

Table 5.1
Major International Climate Meetings and Conferences

1979	First World Climate Conference
1988	World Meteorological Association and United Nations Environmental Program establish the Intergovernmental Panel on Climate Change
1990	Second World Climate Conference
1990	UN General Assembly establishes Intergovernmental Negotiating Committee for the Framework Convention on Climate Change
1992	Signing of Framework Convention on Climate Change at the United Nations Conference on Environment and Development. It provides: *Developed countries*: Reduce greenhouse gases to 1990 levels; Provide cost of compliance of developing countries; *Voluntary compliance encouraged for developing countries*: Cost effectiveness and sinks recognized
1994	Framework Convention on Climate Change (FCCC) enters into force
1995	First Conference of the Parties (FCCC), Berlin, Germany
1996	Second Conference of the Parties (FCCC), Geneva, Switzerland
1997	Third Conference of the Parties (FCCC), Kyoto, Japan
1997	Kyoto Protocol signed view to reducing . . . overall emissions . . . by at least 5 percent below 1990 levels in the . . . period 2008–2012
1998	Fourth Conference of the Parties (FCCC), Buenos Aires, Argentina
1999	Fifth Conference of the Parties (FCCC), Bonn, Germany
2000	Sixth Conference of the Parties (FCCC), Part I, The Hague, Netherlands
2001	Sixth Conference of the Parties (FCCC), Part II, Bonn, Germany
2001	Seventh Conference of the Parties (FCCC), Marrakech, Morocco

Table 5.1
(continued)

2002	Eighth Conference of the Parties (FCCC), New Delhi, India
2003	Ninth Conference of the Parties (FCCC), Milan, Italy
2004	Tenth Conference of the Parties (FCCC), Buenos Aires, Argentina
2005	Kyoto Protocol goes into effect in February
2005	First Meeting of the Parties to the Kyoto Protocol and Eleventh Conference of the Parties (FCCC)
2006	Second Meeting of the Parties to the Kyoto Protocol and Twelfth Conference of the Parties (FCCC), Nairobi, Kenya

The FCCC recognizes that an important part of sustainable development is the conservation and enhancement of *sinks* (processes that remove greenhouse gases and aerosols from the atmosphere) and reservoirs of greenhouse gases. According to the agreement, sinks are "biomass, forests and oceans as well as other terrestrial, coastal and marine ecosystems."

Greenhouse-gas emissions were to be reduced based on currently available scientific knowledge, despite areas of continuing uncertainty:

> The Parties should take precautionary measures to anticipate, prevent or minimize the causes of climate change and mitigate its adverse effects. Where there are threats of serious or irreversible damage, lack of full scientific certainty should not be used as a reason for postponing such measures, taking into account that policies and measures to deal with climate change should be cost-effective so as to ensure global benefits at the lowest possible cost. (FCCC 1992, art. 3)

FCCC nations—those that ratify, approve, accede to, or accept the agreement—promised to meet each year at a conference of the parties (COP) to discuss advances in scientific understanding of climate change and update details regarding implementation. At these conferences, thousands of delegates and observers address the almost innumerable steps required to implement the agreement. Opened for signature in 1992, the FCCC went into effect in 1994. Almost all of the countries and organizations eligible to become parties to the FCCC have done so (189 out of 194), but this level of support for international efforts to address climate change has not been seen since.

The first of the FCCC's conferences of the parties was held in Berlin in 1995, and negotiations were tense. With significant leadership from India, however, a majority of the parties agreed to establish binding reduction targets for developed countries but not for developing countries. At the time, the United States supported the agreement (Carpenter, Chasek, and Wise 1995).

Small island states, the Netherlands, Germany, and Malaysia, among others, argued for stronger reductions in greenhouse-gas emissions. In one session, the representative from Mauritius said "his delegation did not have very much to be proud of, and will leave Berlin with a sense of sadness for having something that is 'half-baked'" (Carpenter et al. 1995). In contrast, some oil-exporting countries said the conference had gone too far. Russia and the United Arab Emirates questioned the scientific basis for action, and Venezuela challenged the conclusion that existing commitments were inadequate. Saudi Arabia, Venezuela, and Kuwait placed reservations on the decisions reached at the first conference (Carpenter et al. 1995).

By the time the parties met for the second formal FCCC conference of the parties in 1996, significant disagreements remained, even though smaller meetings had been held between the two conferences. One of the central points of conflict was whether to use the *Second Assessment Report* of the Intergovernmental Panel on Climate Change (1995) as the basis for negotiating binding agreements on developed countries to reduce greenhouse-gas emissions. Opponents—mainly oil-exporting countries—disagreed with the report's conclusion that "the balance of evidence" indicates that human-caused emissions of greenhouse gases are having a "discernible" influence on global climate change. To the encouragement of some environmentalists, the U.S. insurance industry supported early and substantial reductions of greenhouse-gas emissions to help reduce the risk of severe weather and associated property damage (Carpenter et al. 1996).

A number of other issues also were controversial. The United States had originally opposed legally binding targets, but decided to support them provided that a country could reach its emissions targets by acquiring a right of emission from another entity that has met or does not have to meet a target. Another issue was that developing countries and oil-exporting countries wanted stronger protections against possible economic losses. A number of these nations argued against market-based approaches (such as tradable carbon dioxide permits) because the poor, with perhaps the greatest demand for CO_2 reduction, would be least able to pay to communicate what they need to the market.

Despite the turmoil surrounding the first two FCCC conferences of the parties, several concepts were developed that would be part of the basis for negotiations of the Kyoto Protocol at the third conference of the parties in 1997 (Carpenter et al. 1996):

- Legally binding reductions in greenhouse-gas emissions should be established for industrialized countries,
- Developing countries should be exempt,
- The *Second Assessment Report* should "provide a scientific basis for urgently strengthening action . . . to limit and reduce emissions of greenhouse gases,"

• The IPCC needs to further reduce scientific uncertainties, "in particular regarding socio-economic and environmental impacts on developing countries, including those vulnerable to drought, desertification or sea-level rise."

A minority of countries have maintained strong reservations about FCCC implementation. Many governments believe that current measures are inadequate, and some argue they cannot afford to do more. These positions were raised again during the negotiations at Kyoto in 1997. Nonetheless, the Kyoto Protocol is consistent with the agreements reached earlier by the FCCC countries.

Kyoto Signatories

Countries representing 55 percent of 1990 emissions from the industrialized countries (the Annex 1 countries) had to ratify, approve, accede to, or accept the Kyoto Protocol to bring it into effect. This meant that either the United States or Russia had to accept the agreement, and ninety days after Russia ratified it (on November 18, 2004), the Kyoto Protocol went into effect on February 16, 2005. By then, several additional conferences of the parties had been held where compromises were reached, issues set aside for later consideration, and important programs developed for eventual implementation of the Protocol. The first meeting of the parties (MOP) of the Kyoto Protocol took place simultaneously with a conference of the parties of the FCCC in December 2005 in Montreal, and the parties met (for The Second MOP and twelfth COP) in Nairobi in November 2006.

Between 1997 and November 2004, many of the 124 countries that had already ratified, accepted, approved, or acceded to the Kyoto Protocol began working to comply with their commitments, without waiting for the holdouts.

European Union

The European Union, representing almost 30 percent of 1990 emissions from industrialized countries, led in implementing policies to reduce greenhouse-gas emissions before the Kyoto Protocol went into effect. In a 2003 survey conducted by the European Community, 88 percent of European voters supported taking immediate actions to address climate change (European Commission n.d.).

By signing the Kyoto Protocol, countries in the European Union agreed to reduce their greenhouse-gas emissions by 8 percent of 1990 levels by 2008 to 2012, although targets for individual EU countries vary. The EU Parliament has made this goal legally binding, and a number of regional and national policies—including the creation of a trading system for CO_2 emissions—aim to reach this target. The design of this market has been controversial. In Germany, for example, environmentalists supported an EU-wide market with mandatory compliance by individual companies, while industry groups supported voluntary participation in a market designed to help each country in the EU attain its greenhouse-gas reduction targets (German industry 2001). Other members have revoked or objected to new tax instruments: Finland repealed its carbon tax, Sweden weakened its tax, and France and England have strongly resisted EU-wide carbon taxes (Rabe 2002, 149).

In addition to emissions trading, the European Commission has strengthened energy-efficiency requirements for both residential and nonresidential buildings. In Europe, where buildings consume 40 percent of energy (more than any other part of the economy), energy-efficiency advocates argue that the European Union could exceed its Kyoto Protocol targets through improved insulation, heating, cooling, and lighting technology and like actions—reducing greenhouse-gas emissions by 12.5 percent by 2010 (EU aims to achieve 2001).

The United Kingdom has taken a leading role in addressing climate change. The position of the Blair government is that climate change is an urgent matter, action is needed now to avoid disaster, and policies must encourage investment in science and technology and behavior changes to reduce greenhouse-gas emissions and expand the economy. The UK was the first country to set a target that was more aggressive than the Kyoto Protocol—reducing carbon dioxide emissions by 20 percent by 2010. It has also set a goal of reducing carbon dioxide 60 percent by 2050. To help achieve these goals, the Blair government initiated a series of measures, including a controversial climate-change levy.

Germany and the Netherlands are also taking innovative steps to reduce greenhouse-gas emissions, including greatly increasing their use of renewable energy, particularly offshore wind energy. The Dutch are even controlling the greenhouse gases from their massive flower-producing greenhouses (Verjee 2006). Germany has a Kyoto agreement target of 21 percent below 1990 levels by 2008 to 2012 and has set an additional target of 40 percent reduction of CO_2 by 2020. Germany plans to close down its nineteen nuclear power plants by 2020, although some in the German government have asked whether that action will make achieving the 2020 CO_2-reduction target impossible (Minister doubts 2001). Germany is aggressively developing renewable energy, aiming for about 50 percent by 2050 (Federal Republic of Germany n.d.).

Japan

Japan agreed to cut greenhouse-gas emissions to 6 percent below 1990 levels by 2008 to 2012. In 2002, the government of Japan reported that about 90 percent of its greenhouse-gas emissions came from the combustion of fossil fuels for energy. Japan also reported the highest energy efficiency in the world, so reaching its Kyoto targets will not be easy.

One Japanese policy is to develop new products to help business, individuals, and the government conserve energy. Voluntary action by industry is central. To encourage residential conservation, the government is promoting technology that displays the cost of energy as it is being used in the home (Government of Japan 2002, 72–78).

Japan has also explored the use of a carbon tax, but the idea has not yet become law. Another market-based policy, a limited voluntary greenhouse-gas emissions-trading market among thirty-four participants, has been set up. It allows trading from April 2006 through March 2007 and again for a brief period in June 2007. Participants that do not achieve their voluntary targets must return subsidies awarded to them for greenhouse-gas emissions from Japan's Environment Ministry (Japan for Sustainability 2005).

In the emissions-trading market, businesses may help meet their goals by increasing their forested land (Chasing Kyoto 2004). This option protects biodiversity and water resources and helps to compensate for greenhouse-gas emissions by taking carbon dioxide out of the atmosphere (Ohki 2002).

The Japanese government also advocates switching from fossil fuels to renewable energy and is second only to Germany in its rapidly growing markets for photovoltaic energy generation. In addition, it has called for expanding nuclear energy as a part of its plans for reaching its greenhouse-gas emission targets under the Kyoto Protocol. But as of 2002, public opinion in Japan was opposed to expansion of nuclear energy (Government of Japan 2002, 98).

China

China ranks among the world's top five greenhouse-gas emitters. China has signed the Kyoto Protocol and plans to participate in the aspects of the agreement that apply to developing countries, including updates of national inventories of human-caused greenhouse-gas emissions, programs to help lessen the amount or speed of climate change, and programs to adapt to climate change. The Kyoto Protocol has no mandatory greenhouse-gas reduction targets for developing countries.

Because of China's size, rapid growth, and heavy reliance on coal, investments in China's energy sector during the next few years could have a large impact on climate change for decades to come. China is planning to build more than five hundred new coal power plants by 2012 and is looking to the Kyoto Protocol as well as the Asia Pacific Partnership on Clean Development and Climate (a new group formed by the United States and five other coal-intensive countries) for foreign investment to reduce greenhouse gases from coal. China plans to submit two hundred or more projects for approval under the Kyoto Protocol's clean-development mechanism, which allows developed countries to earn credits by reducing greenhouse-gas emissions in developing countries (Bezlova 2006).

The United States: The Federal Government

The United States had over 36 percent of the 1990 emissions in the industrialized countries (the Annex 1 countries), and its position on climate-change law and policy has shifted and vacillated over the last two decades. Table 5.2 summarizes some of these positions.

In March 2001, the United States surprised the world by reneging on its earlier support of the 1997 Kyoto Protocol. However, as early as July 25, 1997, the U.S. Senate had passed the Byrd-Hagel resolution (Senate Resolution 98) by 95 to 0 to pressure Kyoto negotiators for a global agreement rather than an agreement that bound only industrialized countries. The resolution indicated that Senate support would be withheld if (1) the agreement reached at Kyoto imposed binding limits on industrialized countries without also imposing binding limits on developing countries or (2) the agreement would result in serious harm to the economy of the United States.

The United States signed the Kyoto Protocol in November 1998, but President Bill Clinton did not send it to the Senate for ratification. Although the agreement advanced international efforts to address climate change, it failed to comply with the Byrd-Hagel requirement that developing countries would have to limit greenhouse gases.

Some developing countries had recently joined the Organization for Economic Cooperation and Development (OECD), a status that is relevant for some climate-change commitments, but they had not yet taken responsibility for controlling their greenhouse-gas emissions. Furthermore, some had very low, stable levels of greenhouse-gas emissions, and others, such as China, had rapidly growing emissions. China was and is on course to become the world's largest emitter of greenhouse gases by 2015.

To lower emissions of greenhouse gases, the United States continued to negotiate bilateral agreements with developing countries, including China. Its international efforts, such as

Table 5.2
The U.S. response to climate change

Date	Federal Action
1980s	Gore-Wirth legislative initiatives call for increased funding on climate research and energy conservation and renewable energy.
1987	Congress passes the Global Climate Protection Act, which calls for climate-change policy to be coordinated nationally.
1992	George H. W. Bush reluctantly attends the conference where the United Nation Framework Convention on Climate Change is adopted in Rio de Janeiro.
1992	The United States signs and ratifies the Framework Convention on Climate Change.
1992	Congress passes the Energy Policy Act; it sets rules for restructuring electricity delivery, reducing American dependence on foreign oil, and requiring electricity-generating utilities to report annually on carbon dioxide emissions.
1995	The Clinton administration endorses the Berlin Mandates calling for development of binding emission reductions.
1997	The Byrd-Hale resolution passes 95 to 0 and recommends that no future climate-change agreements be signed without commitments from developing countries.
November 1998	The Kyoto Protocol is signed but is not submitted to Congress.
1998–2000	Bilateral agreements with developing countries on climate-change emissions; participation in Global Environmental Facility projects.
Early 2001	George W. Bush cabinet considers a policy proposal to reduce greenhouse gases and further engage in international meetings.
March 2001	Bush disengages from the Kyoto Protocol.

Table 5.2
(continued)

Date	Federal Action
October 2002	The Bush Justice Department supports a legal challenge to California's program to promote low-emission vehicles.
2002	The Bush administration focuses on carbon-intensity reductions and Voluntary Innovative Sector Initiatives Opportunities Now (VISION).
2003	The McCain-Lieberman bill proposes that greenhouse-gas emissions be reduced to 2000 levels by 2010 and is defeated 56 to 44.
2005	McCain and Lieberman reintroduce their greenhouse-gas bill.
Present	Agencies continue bilateral and regional research.

contributions to an international institution for environmental improvements called the Global Environment Facility, continued, but Clinton considered it premature to submit the Kyoto Protocol to the Senate for ratification because it did not require binding commitments from developing countries.

In the 1990s, the U.S. government faced strong pressure from industry to oppose the Kyoto Protocol. A group named the Global Climate Coalition attended negotiations, contributed to the IPCC scientific assessment documents, provided comments on proposed government programs on climate change, and lobbied Congress. According to this coalition—which included large and small businesses in agriculture, forestry, electric utilities, railroads, transportation, manufacturing, mining, oil, and coal—the Kyoto Protocol was unfair because developing countries were not required to reduce emissions, U.S. growth could be severely hampered if energy prices for consumers increased dramatically as a result of its requirements, and its targets and timetables would gravely harm American families, workers, older people, and children (Global Climate Coalition n.d.; SourceWatch n.d.). The coalition disbanded once it believed that the Bush administration was set to change the U.S. position on the Kyoto Protocol.

In March 2001, President George W. Bush rejected the Kyoto Protocol: "it exempts 80 percent of the world, including major population centers such as China and India, from compliance, and would cause serious harm to the U.S. economy." Bush said that the United States generates more than half of its electricity from coal and that caps on CO_2 emissions would shift electricity generation from coal to natural gas, which would raise energy costs:

> At a time when California has already experienced energy shortages, and other Western states are worried about price and availability of energy . . . we must be very careful not to take actions that could harm consumers. This is especially true given the incomplete state of scientific knowledge of the causes of, and solutions to, global climate change and the lack of commercially available technologies for removing and storing carbon dioxide. (White House 2001)

Californians and Americans in general do not agree with this statement. In April 2001, ABC reported on a survey that indicated that 61 percent of Americans, including 52 percent of Republicans, rejected the arguments made by Bush and said the United States should ratify the Kyoto Protocol. In a 2002 Harris Poll, more than 50 percent of those who had heard of climate change disagreed with Bush's rejection of the Kyoto Protocol.

A bipartisan Senate bill introduced in 2003 proposed that greenhouse-gas emissions be reduced to 2000 levels by 2010. The McCain-Lieberman bill required that a cap be reached through on-site measures or by the trading of greenhouse-gas emission rights. This bill failed by a vote of 55 to 43. Opponents argued that carbon dioxide poses no direct threat to public health and that the McCain-Lieberman requirements would burden families and communities. Supporters said that climate change is real and that problems such as the loss of sea ice in the Arctic would worsen until society reduces greenhouse-gas emissions. In 2005, the senators revised their bill and, in part to attract support from the business community, added hundreds of millions of dollars in subsidies for cleaner electrical energy and nuclear energy (A shift to green 2005).

Rather than reduce total greenhouse-gas emissions, in 2002 the Bush administration adopted the Global Climate Change Initiative, which aims to reduce the ratio of greenhouse-gas emissions to economic output by 18 percent by 2012 through domestic voluntary actions and continued research on climate change. Reaching the target was compared to reducing U.S. greenhouse-gas emissions to 5 percent below 1990 levels by 2008 to 2012. According to the Bush administration, the initiative "ensures that America's workers and citizens of the developing world are not unfairly penalized." In 2002, the administration implied that when stronger scientific justification became available, the United States would stop and even reverse the growth of its greenhouse-gas emissions.

The U.S. position emphasizes research. The federal government provides about $3 billion per year for continuing research on climate change. Support for this was written into law in 1990, and in the president's budgets for later years, including fiscal years 2004 and 2005, federal agencies reported plans for continued research. The *Strategic Plan for U.S. Climate Change Science* (U.S. Climate Change Science Program et al. 2003/2004) states that "the Earth's environmental and human systems are undergoing changes caused by a variety of natural and human-induced causes and . . . technological breakthroughs will be needed to address the long-term challenge of global climate change."

The United States is doing bilateral and regional research with both developing and developed countries. These efforts include a focus on CO_2 reduction. Domestic research aims to reach the 18 percent by 2012 target for reducing greenhouse-gas intensity through use of hydrogen fuels, clean coal, and forecasting, and adaptation to changes in global climate.

Nonetheless, the Pew Center on Global Climate Change, an independent organization whose mission is to provide information and solutions to address global climate change, estimates that the current target for reducing greenhouse-gas intensity will lead to U.S. greenhouse-gas emissions 30 percent above 1990 levels by 2010. Pew argues that a more effective approach would be to augment policies that increase energy efficiency in buildings and appliances. This would make a dent in the source of about one-third of U.S. greenhouse-gas emissions and reduce operating costs associated with electricity (Claussen 2004).

Meanwhile, Congress has taken some initiatives to influence the U.S. position. In June 2005, the Senate passed a resolution stating its intention to require "at some future date, a program of mandatory greenhouse gas limits and incentives." In December 2005, while the first meeting of the parties to the Kyoto Protocol was taking place, twenty-four senators wrote a bipartisan letter to President Bush urging strong United States action at that meeting. The letter mentioned the United States' legal obligation under the FCCC to aid in "preventing dangerous anthropogenic interference with the climate system." With the Democratic majority in the Congress following the November 2006 election, additional initiatives began.

Interest-Group Politics

The Global Climate Coalition is one of the major national environmental groups that have influenced the official U.S. response to climate change over the last decades. Within any country, interest groups affect nations' reactions to climate change and their decisions on entering international agreements. Some OPEC nations have taken very strong positions limiting international regulation of fossil-fuel use, whereas Norway, a large oil exporter, is also highly committed to reducing greenhouse gases at home (Dolsak 2001).

In the United States, the national position on climate change develops within various Congresses and administrations based on contributions from environmental groups, industry groups (including oil, nuclear, and alternative energy), transportation companies, farmers, ideology-based public-interest organizations, labor unions, cities, counties, and states.

Some lobbyists focus specifically on climate change, and others promote general positions on government regulations, some of which influence climate. Some organizations materialize for a particular battle, and others are more or less permanent throughout the environmental wars. The outcome is an "anguished, often moralistic, rhetoric that has polarized national debate and made any semblance of consensus at that level so elusive" (Rabe 2002, 23).

Partisan politics can explain some but not the entire story of the official U.S. position on climate change. Democrats generally have supported official cooperation in international agreements on climate change, and Republicans generally have been skeptical if not downright opposed, but this breakdown does not tell the whole political story. Although Al Gore promoted international responses to the problem when he was vice president (for example, by attending the Kyoto Protocol meeting and by publishing his book *Earth in the Balance*), the Clinton administration backed away from a Gore proposal to set a tax on the use of fossil fuels, and when Congress turned Republican in November 1994, "climate change policy moved even further to the recesses" (Rabe 2002, 12).

Still, a few Republicans have taken vocal positions promoting federal action on climate change. Some interest groups that are critical of government regulations have recognized that tight controls on climate gases might result in profits for certain industries. Other political groups, such as conservative Christians, see climate change as fundamentally an ethical or moral issue. Those that emphasize international affairs hope for liberation from Middle East dominance of the energy industry if oil becomes a less important part of the energy sector.

The States and Cities

Federal policies do not tell the whole story on U.S. climate change. Public opinion is generally supportive of efforts to address climate change, and many state governments and cities have passed laws or adopted policies addressing this issue. In 2003, more than three out of four Americans thought the United States should reduce greenhouse-gas emissions (Leiserowitz 2003; Gallup 2003).

The fifty U.S. states have varying views on climate change. At least sixteen states (including Alabama, Kentucky, South Carolina, and Virginia) have passed resolutions asking the federal government to reject the Kyoto Protocol (Rabe 2002, 20), while others have taken actions that go far beyond its commitments. In 2003, the attorneys general of Connecticut, Maine, and Massachusetts filed a federal lawsuit alleging that the U.S.

Environmental Protection Agency (EPA) under the Bush administration had failed to implement the Clean Air Act to include regulation of carbon dioxide. Under the Clean Air Act, the EPA is required to review and, where appropriate, revise regulations based on scientific information about the environmental health effects of a proposed targeted pollutant. The 2003 lawsuit asserts that carbon dioxide is a pollutant "that causes global warming with its attendant adverse health and environmental impacts" and calls on the Bush administration to revise the Clean Air Act to include regulation of carbon dioxide (Rabe 2002, 164–65). In April 2007, the United States Supreme Court, by a vote of 5 to 4, ruled in favor of the states.

In 2004, eight state governments and the city of New York filed a lawsuit seeking to hold five companies (with 174 sites in twenty states) responsible for contributing, through greenhouse-gas emissions, to a public nuisance under the common law. The plaintiffs seek to cap the defendants' CO_2 emissions and reduce them by amounts determined by the court each year for ten years. Many states are not waiting for the outcome of this lawsuit to do something about climate change. The EPA reports that as of May 2004, twenty-eight states and Puerto Rico had completed plans to reduce these emissions (U.S. EPA n.d., Action plans). Many of these were expected to reduce greenhouse-gas emissions to 1990 levels by 2010 in the participating states.

More than half of the states have passed laws to reduce greenhouse-gas emissions. States bordering the Atlantic, Pacific, and Great Lakes are leaders. In 2001, the New England governors and eastern Canadian premiers committed to reduce greenhouse-gas emissions to 1990 levels by 2010 and at least 10 percent below 1990 levels by 2020. In addition, they agreed to reduce emissions sufficiently over the long term to eliminate any dangerous threat to the climate, which scientists estimate to be 75 to 85 percent below current levels. Some states have passed laws that create mandatory cap and trade programs, set greenhouse-gas reduction targets, and require reporting and inventorying. In addition, some states have laws that promote the capture and storage of carbon dioxide in trees, underground, and other locations and that strengthen energy efficiency, conservation, and renewable energy in electricity generation and transportation (Pew Center on Global Climate Change 2004).

About a third of the states have significant policies aimed at reducing greenhouse gases. As early as 1989, New Jersey had an executive order that called on state government to lead in decreasing emissions of climate-change gases. In 1998, the state, under Governor Christine Whitman (who later became the U.S. EPA administrator), pledged to reduce emissions by 2005 in a way that would reach the goals of the Kyoto Protocol within a few years. Another Republican governor, George W. Bush, signed the Texas Public Utility Regulatory Act of 1999, which required state utilities to move toward greater use of renewable energy sources with the estimated reduction in carbon dioxide emissions by 1.83 million metric tons a year by 2009 (Rabe 2002, 1). In 2000, Nebraska, with only one dissenting vote in the legislature, passed a program designed to sequester carbon through agricultural practices. Several nearby states followed suit. New Hampshire and Massachusetts have applied a cap on carbon dioxide and other pollutants. Oregon has linked siting of new energy-plant facilities to commitments to reduce climate-change gas emissions. Wisconsin has mandated public disclosure of releases of carbon dioxide on a yearly basis; this applies to all large emission sources. These programs can have measurable impacts: Texas regularly emits more greenhouse gases annually than does France, and Wisconsin surpasses Uzbekistan in emissions (Rabe 2002, 1).

Some states have shown interest in international cooperation. New Jersey would like to develop programs with the Netherlands on climate change; Illinois, with China; Nebraska, with other Kyoto countries; and New England states continue to work with Canadian provinces with which they have formal agreements, independent of Canada's federal government (Rabe 2002, 165). There may be constitutional issues raised by some of these efforts, but the desire remains.

As political scientist Barry Rabe explains, the politics of environmental policy at the state level are quite different from those nationally. This is true for climate change: "Contrary to the kinds of political brawls so common in debates about climate change policy at national and international venues, . . . state-based policymaking has been far less visible and contentious, often cutting across traditional partisan and interest group fissures" (Rabe 2002, 22). Some legislation, like that involving motor vehicles in California, faces industry opposition but still is supported by several different groups—in the California case, environmentalists, Silicon Valley business leaders, and big cities (Rabe 2004, 143): "American states may be emerging as international leaders at the very time the national government continues to be portrayed as an international laggard on global climate change" (Rabe 2004, xiv).

At the state level, there also is room for what political scientists call *policy entrepreneurs* to move climate-change policy. Policy entrepreneurs know where change is possible and have expertise that is not seen as politically motivated: "These entrepreneurs have tailored policies to the political and economic realities of their particular setting and have built coalitions that seem almost unthinkable when weighed against the past decade of federal-level experience" (Rabe 2002, 151). Nonetheless, the activity of the states may provide lessons for the federal government if the United States decides to change its formal policies. In the meantime, the states are learning from each other, and ideas are flowing back and forth through interstate meetings and conferences.

California: Register, Clean, Renew

Climate change generates various responses—indifference, scorn, alarm, advocacy, and action. Climate change will affect families and economies in different ways. In some coastal areas, for example, it will have direct and negative impacts, while elsewhere the effects will be mild or beneficial.

California is working to address climate change on many levels, and many of its environmental policies have become national policy. The size of California's economy and population and the volume of its carbon dioxide emissions indicate that activities to reduce greenhouse-gas emissions in that one state can influence efforts around the globe.

For more than twenty years, California has had policies of energy efficiency, pollution reduction, and renewable energy. These are part of a *no-regrets policy* on climate change: "California's energy policies have been carried out primarily to reduce costs to consumers and air pollution. However, these 'no regrets' policies have set California on a firm path to respond effectively to growing concerns about the effects of human-caused greenhouse gas emissions on the earth's ecosystem."

In September 2000, in the middle of an energy crisis, California began developing a voluntary registry for stationary and mobile emissions of carbon dioxide and other greenhouse gases (California Climate Action Registry n.d.) The California Climate

Action Registry tracks voluntary greenhouse-gas emissions reductions by its members, including businesses, government agencies, and other organizations. Members establish baseline emissions and report annual results.

This approach aims to make reductions economically rewarding. Economic gains are expected from green marketing, waste minimization, greenhouse-gas emissions-trading markets, and the export of new technologies. Companies and cities can influence the design of any future mandatory greenhouse-gas emission policy.

California is also regulating greenhouse-gas emissions from cars and trucks. The state has put substantial resources into cutting automobile emissions regulated by the U.S. Clean Air Act. Prior to 2002, little work had been done to reduce CO_2 emissions from transportation sources, despite the fact that they produce more than half of California's CO_2 emissions. But a 2002 state law (AB1493) requires personal vehicles sold in California in model year 2009 and later to comply with greenhouse-gas emission standards to achieve "the maximum feasible and cost-effective reduction" of greenhouse gases. The law encourages early action by rewarding reductions reported through the California Climate Action Registry. Because each gallon of gasoline burned results in about 20 pounds of CO_2 in vehicle exhaust, environmental improvements from this type of pollution prevention should be substantial.

Also in 2002, California created a Renewables Portfolio Standard, which aims to increase eligible renewable energy from 10 percent to 20 percent of retail electricity sales by 2017. State agencies are working to accelerate this target to 20 percent by 2010.

In early June 2005, Governor Arnold Schwarzenegger signed an executive order on climate change before an audience of business and environmental leaders at the United Nations World Environmental Day conference in San Francisco. The governor declared, "As of today, California is going to be the leader in the fight against global warning. . . . I say the debate is over. We know the science, we see the threat, and the time for action is now" (A shift to green 2005).

Under this executive order, the state will reduce its greenhouse-gas emissions to 2000 levels by 2010 and to 1990 levels by 2020. In addition, it will reduce its greenhouse-gas emissions to 80 percent below 1990 levels by 2050. The governor's plan includes increasing energy efficiency and getting a third of California's electricity from renewable resources.

As New Jersey has attempted to bridge international boundaries with its climate change policy, California in 2006 entered into an agreement with the United Kingdom calling for further research on the economics of climate-change policies and new technologies and looking forward to a possible emissions-trading program between the state and the UK (Schoch and Wilson 2006). The international press coverage of Schwarzenegger's meeting with UK Prime Minister Tony Blair suggests some of the intrigue in domestic and international politics linked to climate change.

These actions are broadly supported by Californians. In July 2004, as noted earlier, nearly three of four California residents believed that immediate steps should be taken to counter the effects of global warming. Furthermore, 87 percent supported doubling the use of renewable energy over the next ten years.

Some cities have taken action on climate change. The mayor of Seattle, Greg Nickels, began a program to promote the goals of the Kyoto Protocol though actions by American cities (City of Seattle 2007).

Participating cities commit to meet or exceed Kyoto Protocol targets in their communities; encourage the state and federal government to meet or exceed a 7% reduction from 1990 emission levels by 2012; and urge the U.S. Congress to pass greenhouse gas reduction legislation. As of December 2006, 333 cities had signed on to the agreement. The agreement's action plan discourages urban sprawl, promotes public transportation and the creation of car-pooling and bicycle lanes, and encourages the use of alternative energy sources.

Voters in Boulder, Colorado, approved a carbon tax in 2006. The tax will be based on usage of kilowatt-hours (City approves "carbon tax" in effort [to] reduce gas emissions 2006). Other cities in Colorado have begun programs to purchase wind and other alternate fuels and to have their ski resorts convert to renewable energy for lifts, hotels, and related buildings (Gumbel 2006).

Other Actions

The private sector has also emerged as a major player in the climate-change story in the United States. Many corporations have begun to reduce the financial costs of controls. The World Resources Institute in 2004 concluded that "proactive work" to measure emissions and minimize the costs of rule compliance could be much less costly than "reacting to events at a later date" (Fialka and Ball 2004). Some utilities have taken early steps. New York's Consolidated Edison saved millions of dollars by eliminating natural gas leaks. And the company is selling wind-generated power. Johnson and Johnson has become the United States' second-largest user of solar panels. Pfizer has a goal of reducing greenhouse-gas emissions by 35 percent per dollar of revenue by 2007. General Electric and Citigroup are also leaders in planning for various strategies of emission limitations (Fialka and Ball 2004). In 2005, General Electric committed to double its research into cleaner technologies from $700 million in 2004 to $1.5 billion in 2010. Relative to a 2004 baseline, General Electric plans to reduce its total greenhouse-gas emissions by 1 percent by the end of 2012, reduce its greenhouse-gas intensity by 30 percent by the end of 2008, and improve energy efficiency by 30 percent by the end of 2012.

The EPA reports that more than one hundred corporations totaling almost 10 percent of the United States' gross domestic product and 10 percent of total U.S. greenhouse-gas emissions have joined its voluntary Climate Leaders group since 2002. In January 2006, Citigroup joined the program and set a goal to reduce its greenhouse-gas emissions on a global basis by 10 percent by 2011. SC Johnson has already achieved its goal to reduce greenhouse-gas emissions per pound of product by 23 percent and achieve an 8 percent absolute reduction in these emissions from 2000 to 2005 (30,000 tons a year). To do so, SC Johnson installed a sort of jet engine that generates electricity from landfill gas. In addition to the electricity, the waste heat from the engine is used in manufacturing. The manufacturing plant that installed the system will be able to cut its electricity and natural gas consumption roughly in half and save more than $2 million dollars a year in energy costs (U.S. EPA 2004).

Individual changes in behavior are also suggested to combat climate change. These are simply lessons of environmental education. While some efforts may seem unlikely and their effects insufficient, in aggregate, they can yield substantial emission reductions. For example, in general, more energy is required to produce packaging and newsprint from raw materials than recycled goods. Similarly, public transportation, carpooling, cycling, telecommuting, and flexible work schedules also reduce the greenhouse-gas emissions per person. The EPA estimates that an average two-person household in the United States emits about 60,000 pounds of carbon dioxide per year, with home energy consumption accounting for about three-fourths of the emissions and cars and pick-up trucks making up much of the rest. Using public transportation or telecommuting to cut driving miles from an average of 225 miles per week to 50 miles per week could reduce a household's emissions by more than 10,000 pounds of carbon dioxide per year (U.S. EPA n.d., EPS's personal greenhouse gas calculator).

Rather than viewing prevention and mitigation of climate change as two separate areas, government, business, and agriculture are beginning to link the two and to identify additional actions that could reduce the risk to life and property. Without sizeable reductions of greenhouse-gas emissions, efforts to address climate change—especially adaptation of existing water controls and coastal infrastructure—will become more difficult and costly (Claussen 2004).

Groups like the Pew Center on Global Climate Change promote and advocate a large number of actions that can influence future climate. Since some greenhouse gases, particularly CO_2, remain in the air a hundred years, actions taken today to slow emissions will be most effective if they are designed to withstand changes in climate. Table 5.3 highlights a number of possible current actions that can reduce emissions and affect climate changes.

Over the past decade, rapid technological advancements and increased political pressures have led to lower costs for renewable-energy options. Many European countries, including

Table 5.3
Current actions that can influence future climate changes

Activity	Possible Effect of Climate Change on Activity	Prevention	Mitigation
Carbon sequestration	Insect infestation of forests due to drought	Manage forests for drought conditions using sustainable forest stewardship methods.	Conduct fire-prevention procedures to avoid and contain CO_2 release from fire-prone dead wood.
Energy efficiency (voluntary/short-term)	Temperature swings reducing the willingness or ability of consumers to conserve energy	Follow conservative building practices, such as window and building orientation, deciduous shading, insulation, and flow-through air circulation.	Provide increased incentives for consumers to conserve during periods of peak demand.
Renewable energy	Hydroelectricity reduced, wind more extreme, biomass uncertain, solar uncertain (more clouds)	Accelerate renewable energy development to help reduce greenhouse-gas emissions.	Conduct research and development to improve operations over a broad range of weather conditions.
Agriculture	Droughts or inundation leading to changes in the timing or type of water problems	Employ "no-till" agricultural techniques. Expand the use of waste methane for energy generation.	Change the type of crops to better reflect timing and availability of water supply.

Transportation	Hazardous driving conditions becoming more frequent (e.g., limited visibility, limited traction, extreme heat)	Expand best practices, such as "smart growth," to reduce sprawl and promote public transportation, telecommuting, alternative fuels, and better fuel efficiency.	Provide automakers with incentives for improving vehicle performance under hazardous conditions.
Land use: coastal	Rising sea levels and severe weather leading to greater erosion and mud slides	Create incentives for setbacks from coastal areas and bluffs. Build and strengthen sea walls and levies. Create areas of high ground where possible.	After severe storms, evoke eminent domain with due compensation, and prevent reconstruction within setback area from coast.
Land use: flood plain	Less snow and more rain leading to greater flood risk	Change the flow through dams with new patterns. Build and expand cisterns, floodplain setbacks, and levies to slow and direct flood waters. Separate storm from sewage drains.	After severe storms, evoke eminent domain with just compensation, and prevent reconstruction within setback area from rivers.
Land use: mountain	Extreme weather events leading to greater erosion risk	Maintain healthy vegetation on slopes to help control erosion.	Increase reservoir dredging to maintain the water-storage capacity of dams.

Denmark and Germany, and many U.S. states have established programs for increasing the use of renewable energy. Hydroelectricity provides about 20 percent of California's electricity. An additional 10 percent of California's energy demands are currently satisfied by geothermal, biomass, small hydropower, and wind renewable sources. Excluding large hydropower, by 2010 California plans to satisfy 20 percent of its electricity demands with renewables and has set a goal of 33 percent by 2020 (California Energy Commission 2004). It is also estimated that California could cut in half its transportation-based carbon emissions by improving the fuel efficiencies of vehicles to an average of 40 miles per gallon.

There is much room for change worldwide, where nonrenewable fossil fuels remain between 80 and 85 percent of the supply of primary energy. From 1990 to 2001, the increase in the renewable resource contribution to energy rose only from 7.6 percent to 8.0 percent of consumption. Yet compared to electricity generated by a conventional pulverized coal plant, wind- and solar-generated electricity produce from 130 to 268 times less CO_2 per kilowatt hour (Pirages and Cousins 2005).

Conclusion

Through international organizations, national governments, businesses, and other interest groups, people are learning more and more about climate change and are responding to this extraordinary international challenge. Despite political activity in many nations that minimizes the costs of action and maximizes the benefits of waiting, international efforts have been undertaken, innovations in technologies and policies have been widespread, and nations, states, and provinces are experimenting with policies that go beyond voluntary reductions.

Through international meetings and through focused and sometimes dramatic journalistic coverage of the causes and effects of climate change, global learning is ongoing. Some scientists have concluded that the present response is not enough, but the world community or at least a large portion of it has proved to be capable of reacting to environmental challenges quickly when the true problem is made clear.

As a top-down international effort goes forward, so do activities on the state and provincial levels, in businesses, and even at the individual level. Experiments are ongoing in Denmark, Germany, Japan, the United Kingdom, California, and New Jersey; at British Petroleum, Consolidated Edison, and General Electric; and in neighborhoods in cities throughout the world. As doubts about the existence of climate change and the human contribution are removed, efforts to slow it down and to mitigate its effects have accelerated.

References

ABC News.com Poll. April 11–15. 2001. 〈http://www.pollingreport.com/enviro.htm〉 (accessed on December 5, 2006).

Bezlova, Antoaneta. 2006. Environment: China sends smoke signals on Kyoto Protocol. InterPress Service News Agency, January 20. 〈http://www.ipsnews.net/news.asp?idnews=31842〉 (accessed December 4, 2006).

Boxall, Bettina. 2006. Bush's grade on environment falls. *Los Angeles Times*, August 4, A-22.

Bruntland, G., ed. 1987. *Our common future: The World Commission on Environment and Development*. Oxford: Oxford University Press.

California Climate Action Registry. n.d. 〈http://www.climateregistry.org〉 (accessed December 4, 2006).

California Energy Commission. 2004. *2004 integrated energy policy report update*. 〈http://www.energy.ca.gov/2004_policy_update〉 (accessed December 4, 2006).

Carpenter, Chad, Pamela Chasek, and Steve Wise. 1995. A summary report on the First Conference of the Parties to the Framework Convention on Climate Change: 28 March–7 April 1995. *Earth Negotiations Bulletin* 12(21). Winnipeg, Canada: International Institute for Sustainable Development. ⟨http://www.iisd.ca/vol12⟩.

Carpenter, Chad, Pamela Chasek, Peter Doran, Emily Gardner, and Daniel Putterman. 1996. Summary report of the Second Conference of the Parties to the Framework Convention on Climate Change: 8–19 July 1996. *Earth Negotiations Bulletin* 12(38). Winnipeg, Canada: International Institute for Sustainable Development. ⟨http://www.iisd.ca/vol12⟩ (accessed December 4, 2006).

Chasing Kyoto Pact obligations: Greenhouse gas emissions trade market set for '05. 2004. *Japan Times*, August 4. ⟨http://www.japantimes.co.jp⟩ (accessed December 4, 2006).

City approves "carbon tax" in effort [to] reduce gas emissions. 2006. *New York Times*, November 18, A52.

City of Seattle. 2007. ⟨http://www.seattle.gov/mayor/climate/default.htm⟩ (accessed December 7, 2006).

Claussen, Eileen. 2004. An effective approach to climate change. *Science* 306(5697): 816. ⟨http://www.sciencemag.org/cgi/content/summary/306/5697/816⟩ (accessed December 4, 2006).

Dolsak, Nives. 2001. Mitigating global climate change: Why are some countries more committed than others? *Policy Studies Journal* 29(3): 414–36.

EU aims to achieve Kyoto through energy efficiency. 2001. Reuters, August 17.

European Commission. n.d. Recent standard Eurobarometer. ⟨http://europa.eu.int/comm/public_opinion/standard_en.htm⟩ (accessed June 24, 2006).

Federal Republic of Germany, Federal Foreign Office. n.d. Facts about Germany: Renewable energy sources. ⟨http://web3.s112.typo3server.com/563.99.html⟩ (accessed June 21, 2006).

Fialka, John J., and Jeffrey Ball. 2004. Companies get ready for greenhouse gas limits. *Wall Street Journal*, October 26, A2.

Gallup Organization. 2003. Gallup poll on the environment. April 21. ⟨http://poll.gallup.com⟩ (accessed December 4, 2003).

German industry slams EU emissions trading plan. 2001. Reuters, August 29.

Global Climate Coalition. n.d. Home page. ⟨http://www.globalclimate.org⟩ (accessed September 19, 2004; no longer active as of March 2006).

Government of Japan. 2002. Japan's Third National Communication under the United Nations Framework Convention on Climate Change. ⟨http://unfccc.int/national_reports/annex_i_natcom/submitted_natcom/items/1395.php⟩ (accessed December 4, 2006).

Gumbel, Andrew. 2006. U.S. direct action: How American cities have bypassed Bush on Kyoto. *Independent/UK*, September 1. ⟨http://blog.360.yahoo.com/blog-krLqPvI1erHnhONZ8KT.dCnbXO-?c.q-1&p-877⟩ (accessed December 7, 2006).

Harris Poll. September 19–23, 2002. ⟨http://www.pollingreport.com/enviro.htm⟩ (accessed December 4, 2006).

Information Unit on Climate Change (IUCC). 1979. *The First World Climate Conference*. Geneva: UNEP.

Intergovernmental Panel on Climate Change (IPCC). 1990. *First Assessment Overview and Policy Maker Summaries and 1992 IPCC Supplement*. Geneva., Switzerland: IPCC.

Intergovernmental Panel on Climate Change (IPCC). 1995. *IPCC Second Assessment Climate Change*. Geneva, Switzerland: IPCC.

Japan for Sustainability. 2005. Information center: Japan's voluntary emissions trading scheme starts. September 25. ⟨http://www.japanfs.org⟩ (accessed December 4, 2006).

Leiserowitz, Anthony. 2003. American opinions on global warming. University of Oregon, June. DS Report No. #554. Eugene, OR: Decision Research.

Minister doubts Germany can meet pollution target. 2001. Reuters, March 31.

Ohki, Hiroshi. 2002. Statement by Mr. Hiroshi Ohki, Minister of the Environment, upon Japan's acceptance of the Kyoto Protocol. June 4. ⟨http://www.env.go.jp/en/earth/cc/020604.html⟩ (accessed December 4, 2006).

Pew Center on Global Climate Change. 2004. *Climate change activities in the United States: 2004 update*. ⟨http://www.pewclimate.org⟩ (accessed December 4, 2006).

Pirages, Dennis, and Ken Cousins. 2005. *From resource scarcity to ecological security: Exploring new limits to growth*. Cambridge, MA: MIT Press.

Rabe, Barry G. 2002. Greenhouse and statehouse: The evolving state government role in climate change. Pew Center on Global Climate Change, Washington, DC.

Rabe, Barry G. 2004. *Statehouse and greenhouse: The emerging politics of American climate change policy.* Washington, DC: Brookings Institute Press.

Schoch, Deborah, and Janet Wilson. 2006. Governor, Blair reach environmental accord. *Los Angeles Times*, August 1, B-4.

A Shift to Green. 2005. *Los Angeles Times*, June 12, C1, C5.

SourceWatch. n.d. Global climate coalition. ⟨http://www.sourcewatch.org/index.php?title=Global_Climate_Coalition⟩ (accessed June 24, 2006).

U.S. Climate Change Science Program and U.S. National Science and Technology Council Subcommittee on Global Change Research. 2003|2004. *Strategic plan for U.S. climate change science.* Final Report. July. Updated May 19, 2004. ⟨http://www.climatescience.gov/Library/stratplan2003/final⟩ (accessed December 4, 2006).

U.S. Environmental Protection Agency (EPA). 2004. Climate leaders success stories: Innovative approaches to reducing greenhouse gas emissions. EPA-430-K-04-010. June. ⟨http://www.epa.gov/climateleaders/docs/cl_casestudy.pdf⟩ (accessed December 4, 2006).

U.S. Environmental Protection Agency (EPA). n.d. Action plans. ⟨http://yosemite.epa.gov/OAR/globalwarming.nsf/content/ActionsStateActionPlans.html⟩ (accessed December 4, 2006).

U.S. Environmental Protection Agency (EPA). n.d. EPA's personal greenhouse gas calculator. ⟨http://yosemite.epa.gov/OAR/globalwarming.nsf/content/ResourceCenterToolsGHGCalculator.html⟩ (accessed December 7, 2006).

United Nations Framework Convention on Climate Change (FCCC). 1992. ⟨http://unfccc.int⟩ (accessed December 4, 2006).

Verjee, Neelam. 2006. Dutch growers seek new form of flower power. *Times of London*, June 3, 13. ⟨http://www.timesonline.co.uk⟩ (accessed December 4, 2006).

White House. 2001. Press release: Text of a letter from President George W. Bush to Senators Hagel, Helms, Craig, and Roberts. Washington, DC, March 13. ⟨http://www.whitehouse.gov/news/releases/2001/03/20010314.html⟩ (accessed December 4, 2006).

6

Climate Change as News: Challenges in Communicating Environmental Science

Andrew C. Revkin

A few decades ago, anyone with a notepad or camera could have looked almost anywhere and chronicled a vivid trail of environmental despoliation and disregard. Only a few journalists and authors, to their credit, were able to recognize a looming disaster hiding in plain sight. But at least it was in plain sight. Now, the nature of environmental news is often profoundly different. Biologists these days are more apt to talk about ecosystem integrity than the problems facing eagles or some other individual charismatic species. The subject of sprawl is as diffuse and diverse as the landscapes it encompasses. Concerns about air pollution have migrated—from the choking plumes of old to the smallest of particles that penetrate deep in the lungs and to the invisible heat-trapping greenhouse gases linked to global warming, led by innocuous carbon dioxide, the bubbles in beer. Even though scientists say the main cause of recent warming is smokestack and tailpipe emissions, projections of the pace and ramifications of future climate changes remain as murky as the mix of clouds, particles, and gases that determine how much sunlight reaches the earth and how much heat radiates back into space—the balance that sets the global thermostat.

The challenges encountered in meaningfully translating such issues for the public today are enormous for a host of reasons. Some relate to the subtlety or complexity of the pollution and ecological issues that remain after glaring problems have been addressed. Others relate to effective, well-financed efforts by some industries and groups that oppose pollution restrictions to amplify the uncertainties in environmental science and exploit the tendency of journalists to seek two sides to any issue. This approach can effectively perpetuate confusion, contention, and ultimately public disengagement and inaction.

On the other side of the debate, environmental groups are not innocent in this regard. In some cases, they have focused media attention on their favored issues by going beyond the data and magnifying the risks of, say, cancer or abrupt climate change. Some scientists, expressing frustration with the public's indifference to long-term threats, have stepped outside their areas of expertise and portrayed warming as a real-time catastrophe.

The rhetoric swelled in the spring of 2006 as documentary films, books, and magazine cover stories endeavored to directly link the outbreak of hurricanes, and particularly the ferocity of Hurricane Katrina, to the slow buildup of heat in the world's oceans from human activities. *Time* magazine proclaimed on April 3 "Be worried. Be very worried" (Kluger et al. 2006). A trailer for *An Inconvenient Truth*, the film that documents former Vice President Al Gore's peripatetic multimedia climate campaign, called it "the most terrifying film you will ever see."

Many climate experts said that while there was a growing likelihood that humans were helping shape storm patterns and the like, the inherent variability and complexity of the climate system guaranteed that drawing any straight lines was impossible. On hurricanes, for example, even some of the scientists who claim to have found a relationship between rising hurricane intensity and human-caused warming said that no one could point (with any credibility) to this relationship affecting a particular storm or season.

Critics of those who proclaimed the dawn of a real-time man-made climate catastrophe lashed out. In an opinion piece in the *Wall Street Journal*, Richard S. Lindzen (2006), a climatologist at the Massachusetts Institute of Technology who has long disputed the dominant view that humans could dangerously warm the climate, labeled some calamitous claims "lies" and derided what he called an "alarmist gale."

Between the depictions of global warming as an unfolding catastrophe and as a nonevent lies what appears to be the dominant and still troubling view: that the buildup of carbon dioxide and other long-lived greenhouse gases poses a sufficient risk of profound and largely irreversible transformations of climate and coastlines to warrant prompt action to limit future harms. That view was clearly articulated by eleven national academies of science, including the U.S. National Academy, in a letter to world leaders in 2005.

Many experts explain that it is urgent to act promptly to curb emissions and limit future risks. In fact, because of population growth and increased energy use in developing countries, even the most optimistic scenarios project that concentrations of greenhouse gases will continue to climb throughout the first half of the twenty-first century.

The problem is that the processes that winnow and shape the news have a hard time handling the global-warming issue in an effective way. The media seem either to overplay a sense of imminent calamity or to ignore the issue altogether because it is not black and white or on a time scale that feels like news. This approach leaves society like a ship at anchor swinging cyclically with the tide and not going anywhere.

What is lost in the swings of media coverage is a century of study and evidence that supports the keystone findings: human-generated heat-trapping gases are holding in heat, and the ongoing buildup of this greenhouse-gas blanket adds to warming, shrinks the world's frozen zones, raises seas, and shifts climate patterns.

Certainly, the disinformation generated on both sides of the issue can trip up even earnest, skilled journalists. And the complexity of climate science and policy questions poses a huge challenge in media that are constrained by deadlines and a limited supply of column inches or newscast minutes. Another hurdle is the persistent lack of basic scientific literacy on the part of the public. Nonetheless, some of the biggest impediments to effective climate coverage seem to lie not out in the examined world but back in the newsroom and in the nature of news itself. Overcoming these impediments is a persistent and daunting task. No one should expect to pick up a daily paper anytime soon and read a headline that takes climate science across some threshold of definitiveness that will suddenly trigger public agitation and policy action—and if such a story does appear, it should be looked at skeptically.

A Legacy of Calamity

A little reflection is useful. Most journalists of my generation were raised in an age of imminent calamity. Cold-war "duck-and-cover" exercises regularly sent us to school basements. The prospect of silent springs hung in the wind. We grew up in a landscape where environmental problems were easy to identify. The shores of the Hudson River, for example, were coated with adhesives, dyes, and paint, depending on which riverfront factory was nearest, and the entire river was a repository for human waste, making most sections unswimmable. Across the United States, smokestacks were unfiltered. Gasoline was leaded. Los Angeles air was beige.

Then things began to change. New words crept into the popular lexicon—*smog*, *acid rain*, *toxic waste*. At the same time, citizens gained a sense of empowerment as popular protests shortened a war. A new target was pollution. Earth Day was something new and vital, not an anachronistic notion. Republican administrations and bipartisan Congresses created laws and agencies aimed at restoring air and water quality and protecting wildlife. And remarkably, those laws began to work.

Still, through the 1980s the prime environmental issues of the day—and thus in the news—continued to revolve around iconic incidents that were catastrophic in nature. First came Love Canal, quickly followed by Superfund cleanup laws. Then came Bhopal, which generated the first right-to-know laws granting communities information about the chemicals stored and emitted by nearby businesses. Chernobyl illustrated the perils that were only hinted at by Three Mile Island. The grounding of the *Exxon Valdez* powerfully illustrated the ecological risks of extracting and shipping oil in pristine places. Debates about wildlife conservation generally focused on high-profile species like the spotted owl or whales, and gripping stories in which a charismatic creature was a target of developers or insatiable industries presented simplistic views of reality.

In the late 1980s, the world began to focus on the harm caused by burning in the Amazon and other tropical forests. Forest destruction was made personal and relevant to citizens of the industrialized world when the forests were portrayed as the "lungs of the world" or our "medicine chest"—not because scientists suddenly found a way to describe the extraordinary biological diversity of rain forests and the role they play in the global climate system.

Indeed, the first sustained media coverage of global warming was spawned not by a growing recognition that long-lived emissions from industrial smokestacks and tailpipes could alter the climate. Instead, it began when the American public experienced a record hot summer in 1988 just as satellites and the space shuttle were transmitting images of the thousands of fires burning across the Amazon basin. The burning season in the rain forests was unleashing torrents of carbon dioxide that were perceived as directly perilous to us, so we paid attention. These days, deforestation in the tropics is once again a distant regional issue and has faded to near obscurity in the press—resurging only briefly when someone prominent is gunned down there, like the American nun Sister Dorothy Stang in 2005.

Nuclear Winter, Nuclear Autumn

My first stories about the atmosphere and climate came a few years before the scorching greenhouse summer of 1988 and focused on the inverse of global warming—nuclear winter. Here was a ready-made news story. Prominent scientist-communicators, most notably Carl Sagan and Paul Ehrlich, calculated that anyone surviving a nuclear war might perish in the months of cold and dark that followed as the smoke-veiled sky chilled the earth and devastated agriculture and ecosystems. As the scientists met with Pope John Paul II and the theory made the covers of major magazines, the scenario brought new pressure on leaders to find a way to end the cold war. Within a couple of years, however, fresh scientific analysis showed that the aftermath of nuclear war might be more like a nuclear autumn (to use a phrase coined by Stephen Schneider and Starley Thompson, climate scientists who independently assessed the question). A prediction of nuclear winter was dramatic, dangerous, and novel news. Nuclear autumn was not news, and the double-doomsday scenario quickly faded.

In the meantime, global warming began to build and ebb as a story, always building a bit more with each cycle. If there is one barometer that can help a society gauge whether a problem is real, it is longevity. Unlike concerns about nuclear winter and despite challenges by antiregulatory lobbyists and skeptical scientists, concerns about climate change have not diminished. Instead, evidence of the link and its potential dangers has built relentlessly, as is deftly charted in *The Discovery of Global Warming* by Spencer R. Weart (2003), a historian at the American Institute of Physics.

An Ozone Hole over Antarctica

In the late 1980s, there was a sense of the new about the greenhouse effect, even though scientists had been positing since the 1890s that heat-trapping gases, particularly carbon dioxide released by burning coal and other focal fuels could raise global temperatures. A combination of observations and computer simulations seemed finally to be giving a face to theory, which made it easy to sell as a cover story in *Time* magazine or to *Science Digest*, *Discover*, the *Washington Post*, or the *New York Times*. At that time, there was also a newly perceived global atmospheric threat—the damage to the ozone layer from chlorofluorocarbons (CFCs) and other synthetic compounds—and an international solution in a treaty that banned the chemicals.

But eliminating a handful of chemicals produced by a handful of companies is a very different challenge than eliminating emissions from almost every activity of modern life—from turning on a lamp to driving a car. Another difference between global warming and ozone damage was the iconic nature of the ozone problem. It was an issue with an emblem—the stark, seasonal "hole" that was discovered in the protective atmospheric veil over Antarctica. If a picture is worth a thousand words, a satellite image of a giant purple bruiselike gap in the planet's radiation shield must be worth 10,000. Indeed, according to many surveys, the ozone hole still resonates in the popular imagination—incorrectly—as a cause of global warming simply because it is so memorable and has something to do with the changing atmosphere. The ozone hole also resonated with the public because it was directly linked with an issue that concerns everyone—their health—through the possible risk of increased rates of skin cancer. There, too, global warming is different. Some of the least understood impacts of warming are the possible connections to health problems, like patterns of tropical disease and the frequency of smoggy days, as the National Academies of Science concluded in 2001.

Still, human contributions to the greenhouse effect have remained a perennial issue. Specialized reporters have tracked the developments in climate science and the policy debates over the implications of that science. Tracking scientific progress has become somewhat akin to the old art of Kremlinology—sifting for subtle shifts of language showing that vexing questions are being resolved. Every five years or so, fresh hints emerge from the Intergovernmental Panel on Climate Change (IPCC), the

United Nations scientific body charged with assessing the state of understanding of the problem. The group has sought to be as concrete as possible in its findings, giving quantitative weight to words and phrases such as "likely" and "very likely." That metric has helped the media meaningfully explain the incremental improvements in scientific understanding of the causes and consequences of warming.

The other vital component of the assessment process has been the use of scenarios to depict how certain societal behaviors, particularly energy use, might affect the pace and extent of climate shifts over the course of the century. For the public, this practice provides boundaries for outcomes and a means of judging what kind of response is the most reasonable.

But the incremental nature of climate research and its uncertain scenarios will continue to make the issue of global warming incompatible with the news process. Indeed, global warming remains the antithesis of what is traditionally defined as news. Its intricacies, which often involve overlapping disciplines, confuse scientists, citizens, and reporters—even though its effects will be widespread, both in geography and across time. Journalism craves the concrete, the known, the here and now and is repelled by conditionality, distance, and the future.

If ever there was a moment for a page-one story on climate, for example, it came in October 2000, when a scientist sent me a final draft of the summary for policymakers from the IPCC's third climate assessment, due out early in 2001 (Revkin 2000). For the first time, nearly all of the caveats were gone, and there was a firm statement that "most" (meaning more than half) of the warming trend since 1950 was probably due to the human-caused buildup of greenhouse gases. To me, that was a profound turning point, and I wrote my story that way:

> Greenhouse gases produced mainly by the burning of fossil fuels are altering the atmosphere in ways that affect earth's climate, and it is likely that they have "contributed substantially to the observed warming over the last 50 years," an international panel of climate scientists has concluded. The panel said temperatures could go higher than previously predicted if emissions are not curtailed.
>
> This represents a significant shift in tone—from couched to relatively confident—for the panel of hundreds of scientists, the Intergovernmental Panel on Climate Change, which issued two previous assessments of the research into global warming theory, in 1995 and 1990. (Revkin 2000)

To the *New York Times*, this was just another news story, and it was outcompeted for the front page by presidential politics, the breakup of AT&T, the overthrow of a junta in Ivory Coast, a study on the value of defribillators in public places, and a decision by Hillary Clinton to return some campaign contributions from a Muslim group. Reporters, scientists, and the public can take steps to improve this situation. The first one is simply to anticipate the hurdles that can create trouble when the news media and climate science mix.

The Tyranny of News

A fundamental impediment to coverage of today's top environmental issues is the nature of news. News is almost always something that happens that makes the world different today. A war starts. A tsunami strikes. In contrast, most of the big environmental themes of this century concern phenomena that are complicated, diffuse, and poorly understood, with harms spread over time and space. Runoff from parking lots, gas stations, and driveways invisibly puts the equivalent of one and a half *Exxon Valdez* loads of petroleum into coastal ecosystems each year, the National Research Council (2003) recently found. But try getting a photo of that or finding a way to make an editor understand its implications. A journalism professor of mine once spoke of the "MEGO" factor: "my eyes glaze over." I've seen that look come over more than a few editors in my years of pitching stories on climate.

Climate change is the poster child of twenty-first-century environmental issues. Many experts say that it will be a defining ecological and socioeconomic problem in a generation or two and actions must be taken now to avert a huge increase in emissions linked to warming as economies in developing countries expand. But you will never see a headline in a major paper reading "Global Warming Strikes: Crops Wither, Coasts Flood, Species Vanish." All of those things may happen in plain sight in coming decades, but they will occur so dispersed in time and geography that they will not constitute news as we know it.

Most changes in the landscape and developments in climate science are by nature incremental. Even as science clarifies, it also remains laden with statistical analyses, including broad *error bars*. In the newsrooms I know, the adjective *incremental* in a story is certain death for any front-page prospects, yet it is the defining characteristic of most environmental research. Editors crave certainty: hedging and caveats are red flags that immediately diminish the newsworthiness of a story.

In fact, reporters and editors are sometimes tempted to play up the juiciest—and often least certain—facet of some environmental development, particularly in the late afternoon as everyone in the newsroom sifts for the "front-page thought." They do so at their peril and at the risk of engendering even more cynicism and uncertainty in the minds of readers about the value of the media—especially when one month later the news shifts in a new direction. As a reporter, it can be hard to turn off one's news instinct and insist that a story is not "frontable" or that it deserves three hundred words and not eight hundred, but it is possible—kind of like training yourself to reach for an apple when you crave a cookie.

Scientists have gotten into trouble for doing the same kind of thing. Over and over, I meet scientists who despair that issues they see as vital, like climate change or diminishing biological diversity, are not receiving adequate attention. They feel that they "get it" and the rest of the world does not. When talking to the media, some have been tempted to push beyond what the science supports—focusing on the high end of projections of global temperatures in 2100 or highlighting the scarier scenarios for emissions of greenhouse gases. Recently, a few scientists and environmental groups linked Florida's devastating 2004 and 2005 hurricane seasons to warming, even though the inherent variability in hurricane frequency and targets precludes any such link without a host of caveats and scientific projections call only for slight intensification of tropical storms late in the century, not greater numbers.

The coverage linking these storms to warming oceans resulted in a backlash when some hurricane experts disputed the assertions made to the media. Some statements made to the press about climate and hurricanes were made by climatologists who lacked expertise in the conditions generating these great storms. As a result, in late 2004 one federal hurricane expert, Christopher Landsea, withdrew in protest from the climate-review process at the Intergovernmental Panel on Climate Change, leading to stories on a dispute over climate science. The result was probably more public confusion and cynicism about what is going on.

This tendency of everyone, from scientists to reporters, to focus on the most provocative element when climate becomes news backfired in a very big way in August 2000. A science reporter for the *New York Times* wrote that a couple of scientists on a tourist icebreaker cruise in the Arctic had seen a large patch of open water at the North Pole, possibly the first such occurrence in thousands of years. Better yet, there were pictures. In an interview, one of the scientists ascribed the open water to global warming, and on a quiet summer weekend, the story popped to the top of the front page (Wilford 2000). Finally, the climate-change issue seemed to be behaving like a news story. It was vivid and dramatic, implying that profound changes were afoot. Television reports and political cartoonists quickly followed up with items on the loss of Santa's summer residence.

Unfortunately, the story was incorrect. Calling a few independent experts might have helped the reporter to avoid trouble. Although vast regions in the Arctic may soon be open water in summer and sometime late this century perhaps a blue ocean will exist at that end of the Earth, the sighting in 2000 was unremarkable. Floating sea ice is always a maze of puzzle pieces and open areas. Society would have to wait for its global warming wakeup call. Since 2000, the science has steadily pointed to the ever-growing summertime retreat of Arctic sea ice as an early indicator of human-driven warming. But it remains a subtle process, laden with uncertainty.

After covering climate for over twenty years, my sense is that there will be no single new finding that will generate headlines that galvanize public action and political pressure. Even extreme climate anomalies, such as a decade-long superdrought in the West, could never be shown to be definitively caused by human-driven warming.

The Tyranny of Balance

Journalism has long relied on the age-old method of finding a yea-sayer and a nay-sayer to frame any issue from abortion to zoning. It is an easy way for reporters to show they have no bias. But when dealing with a complicated environmental issue, this method is also an easy way to perpetuate confusion in readers' minds about issues and about the media's purpose. When this format is overused, it tends to highlight the opinions of people at the polarized edges of a debate instead of in the much grayer middle where consensus generally lies. The following maxim illustrates the weakness of this technique: "For every PhD, there is an equal and opposite PhD." The practice also tends to focus attention on a handful of telegenic or quotable people working in the field who are not necessarily the greatest authorities. There are exceptions, but over the years I have learned to be skeptical of scientists who are adept at speaking in sound bites.

One solution to the tyranny of balance is for writers to cultivate scientists in various realms—chemistry, climatology, oceanography—whose expertise and lack of investment in a particular bias are well established. These people can operate as guides more than as sources to quote in a story. Another way to avoid the pitfall of false balance is to focus on research published in peer-reviewed journals rather than that announced in press releases. Peer review, as scientists know all too well, is a highly imperfect process. But it provides an initial quality-control test for new findings that advance understanding of an issue.

The norm of journalistic balance has been exploited by opponents of emissions curbs. Starting in the late 1990s, big companies whose profits were tied to fossil fuels recognized they could use this journalistic practice to amplify the inherent uncertainties in climate projections and thus potentially delay cuts in emissions from burning those fuels. Perhaps the most glaring evidence of this strategy was a long memo written by Joe Walker, who worked in public relations at the American

Petroleum Industry, that surfaced in 1998. According to this "Global Climate Science Communications Action Plan," first revealed by my colleague John Cushman at the *New York Times*, "Victory will be achieved when uncertainties in climate science become part of the conventional wisdom" for "average citizens" and "the media" (Cushman 1998). The action plan called for scientists to be recruited, be given media training, highlight the questions about climate, and downplay evidence pointing to dangers. Since then, industry-funded groups have used the media's tradition of quoting people with competing views to convey a state of confusion even as consensus on warming has built.

A recent analysis of twenty years of newspaper coverage of global warming, including articles in the *New York Times*, showed how the norm of journalistic balance actually introduced a bias into coverage of climate change. Researchers from the University of California at Santa Cruz and American University tracked stories that portrayed science as being deadlocked over human-caused warming, being skeptical of it, or agreeing it was occurring. While the shift toward consensus was clearly seen in periodic assessments by the Intergovernmental Panel on Climate Change, the coverage lagged significantly and tended to portray the science as not settled (Boykoff and Boykoff 2004).

One practice that can improve coverage of climate and similar issues is what I call "truth in labeling." Reporters should discern and describe the motivations of the people cited in a story. If a meteorologist is also a senior fellow at the Marshall Institute, an industry-funded think tank that opposes many environmental regulations, then the journalist's responsibility is to know that connection and to mention it. Such a voice can have a place in a story focused on the policy debate, for example, but not in a story where the only questions are about science. The same would go for a biologist working for the World Wildlife Fund.

Another effective approach is to listen carefully to the facts embedded in what someone is saying, regardless of that person's affiliations. For a 2003 story on the politicization of climate science, for example, I interviewed Patrick J. Michaels, a University of Virginia climatologist and outspoken critic of the mainstream view that human-caused warming is dangerous. While laying out his argument against that view, he said he had recently calculated that the most likely warming in the twenty-first century would be just 1.5°C (2.7°F). Later, I realized that Michaels—a prime skeptic who received income through his affiliation with the Cato Institute, an antiregulatory group that was supported substantially by energy companies—had essentially entered the mainstream. His predicted warming was more than two and a half times the twentieth-century warming and within the range projected by the IPCC.

None of this comes easily, in part because of two more hurdles that constrain a reporter's ability to characterize what is being said in a story.

The Twin Tyrannies of Time and Space

I came to newspapers after writing magazine stories and books and at first was petrified about filing on a daily deadline. One of my editors, hovering over my shoulder and alluding to the stately pace of other forms of publication, while daylight ebbed, gently put it this way: "Revkin, this ain't no seed catalog." Through the ensuing years, I adapted to the rhythm of the daily deadline but also to the reality of its limitations. On an issue like the environment, I understood why the crutch of "on the one hand" was so popular: there is often simply no time to canvass experts. I grew to understand why stories tend sometimes to read like a cartoon version of the world: there is just no time to do better.

And then there is the question of space. Science is one of the few realms where reporters essentially have to presume the reader has no familiarity at all with the basics, particularly something as complicated as climate science. Just about anyone in America knows the rules of politics, business, baseball, and other subjects in the news. But studies of scientific literacy show that most people know little about atoms, viruses, or the atmosphere. So a lot of extra explication somehow has to fit into the same amount of space devoted to a story on a stock split, a primary vote, or a ball game—and it doesn't. Stories about global warming are not granted a few hundred extra words because it is harder than other subjects.

The shrinking of a climate story that is competing on a page with national or foreign developments is as predictable as the retreat of mountain glaciers in this century. But the material that is cut matters to researchers and to those who want to convey the real state of understanding: the caveats, the couching, the words like *may* and *could*, the new questions that emerge with every answer. Labeling ideally should be there to characterize the various voices in a story.

The only solution is to educate editors as much as possible about the importance of context and precision in such stories. That fight is getting more difficult as the media feel more pressure to generate profits and attract readers. More and more, the limited "news hole" reserved for science in newspapers is being filled with stories on subjects most likely to boost circulation, like fitness, autism, diet, and cancer. That leaves ever fewer columns for basic science or research on looming risks like climate change.

Heat versus Light

One of the most difficult challenges in covering the environment is finding the appropriate way to ensure a different kind of balance—between the potent "heat" generated by emotional content and the "light" of science and statistics. Consider a cancer cluster. A reporter constructing a story has various puzzle pieces to connect. Some paragraphs or images brim with the emotional power of the grief of a mother who lost a child to leukemia in a suburb where industrial effluent once tainted the water. A dry section lays out the cold statistical reality of epidemiology, which might never be able to determine if contamination caused the cancer. No matter how one builds such a story, it may be impossible for the reader to come away with anything other than the conviction that contamination killed.

In the climate arena, substitute drowning polar bears or displaced Arctic cultures for cancer-stricken children, and you have the same dynamic at work. It is vital to explore how a warming climate affects ecosystems and people. But this tactic can backfire if a story downplays the uncertainties surrounding unusual climate events or if it portrays everything unusual in the world today as driven by human-caused warming.

It is my impression that the European press, which gives more attention than American media do to climate, has also been more apt to play up hot content and minimize the cooler elements that might deflate a story's sense of drama. This approach caught hold in the United States after Hurricane Katrina, to the extent that the poster for *An Inconvenient Truth* showed a plume from a smokestack merging with a swirling satellite image of a hurricane.

This tactic makes for powerful headlines and gripping TV and magazine images, but are media that adopt this approach doing their job? By the metric of the newsroom, the answer is probably yes. Pushing the limits is a reporter's duty. Finding the one element that's new and implies malfeasance or peril is the key to getting on the front page.

I hope that my own work and that of others will try to refine purely news-driven instincts, to understand and convey the tentative nature of new scientific knowledge, and to retain at least some shades of gray in all that black and white. We also need to drive home that once a core body of understanding has accumulated over decades on an issue—as is the case with human-forced climate change—society can use it as a foundation for policies and choices.

The Great Divide

Journalists dealing with global warming and similar issues would do well to focus on the points of deep consensus, generate stories containing voices that illuminate instead of confuse, convey the complex without putting readers (or editors) to sleep, and cast science in its role as a signpost pointing toward possible futures, not as a font of crystalline answers.

The only way to accomplish this is for reporters to become more familiar with scientists and the ways of science. This requires using those rare quiet moments between breaking-news days to talk to climate modelers, ecologists, or oceanographers who are not on the spot because their university has just issued a press release. By getting a better feel for the breakthrough and setback rhythms of research, a reporter is less likely to forget that on any particular day the state of knowledge about endocrine disruptors, PCBs, or climate is temporary. Readers will gain the resolve to act in the face of uncertainty once they absorb that some uncertainty is the norm, not a temporary state that will give way to magical clarity sometime in the future.

There is another reason to do this. Just as the public has become cynical about the value of news, many scientists have become cynical and fearful about journalism. Some of this is their own fault. When I was at a meeting in Irvine, California, on building better bridges between science and the public, one researcher stood up to recount her personal "horror story" about how a reporter misrepresented her statements and got everything wrong. I asked her if she had called the reporter or newspaper to fix the errors and begin a dialogue about preventing future ones. She had not even considered doing so.

Cynical unconcern for the presumed failings of journalism in part prolonged the career of the disgraced former *New York Times* reporter Jayson Blair. Few of the people who identified falsehoods in his stories called the paper to correct them. The interactions between sources, journalists, and readers ideally should take on more of the characteristics of a conversation. The communication of news cannot remain effective if it is a monologue.

The more scientists and journalists talk, the more likely it is that the public—through the media—will appreciate what science can (and cannot) offer as society grapples with difficult questions about how to invest scarce resources. An intensified dialogue of this sort is becoming ever more important as science and technology increasingly underpin daily life and the progress of modern civilization.

Given the enormous consequences and irreversible losses from global warming should the worst projections play out, the time for improving the flow of information on this subject is clearly now.

References

Boykoff, J. M., and M. T. Boykoff. 2004. Balance as bias: Global warming and the U.S. prestige press. *Global Environmental Change* 14: 125–36.

Cushman, John H., Jr. 1998. Industrial group battles climate treaty. *New York Times*, April 26, pA1.

Gore, Al. 2006. *An inconvenient truth: The planetary emergency of global warming and what we can do about it*. Emmaus, PA: Rodale Press.

Kluger, Jeffrey, et al. 2006. Be Worried. Be Very Worried. Earth at The Tipping Point. *Time*, April 3, 24–54.

Lindzen, Richard. 2006. Climate of fear. *Wall Street Journal*, April 12, A14.

National Academies of Science. 2001. *Under the weather: Climate ecosystem and infectious disease*. Washington, DC: National Academies Press.

National Research Council, Committee on Oil in the Sea. 2003. *Oil in the Sea III: Inputs, fates, and effects*. Washington, DC: National Academies Press.

Revkin, Andrew C. 2000. A shift in stance on global warming theory. *New York Times*, October 26, A22.

Walker, Joe. 1998. Memo, American Petroleum Industry. ⟨www.climatesciencewatch.org/index.php/csw/details/cei_tv-sp⟩ (accessed December 7, 2006).

Weart, Spencer R. 2003. *The discovery of global warming*. Cambridge, MA: Harvard University Press.

Wilford, John Noble. 2000. Ages-old icecap at North Pole is now liquid, scientists find. *New York Times*, August 19, A1.

7

Climate Change and Human Security

Richard A. Matthew

The social effects of climate change and of many other forms of environmental change are varied and hard to predict.[1] This is in large measure because a complex array of interactive variables and nonlinear relationships shape social outcomes. As some parts of the planet are transformed into desert and others disappear under a flood of algae-rich sea water, still other areas will become newly amenable to settlement, agriculture, and oil exploration. As climate change gradually or abruptly reconfigures the planet's geography, living systems—including human communities—will differ significantly in their capacity to adapt, escape, or profit. Jeffrey Sachs (2005) has linked climate change to violent conflicts in Sudan and elsewhere. But *The Observer* (London) has commented on "the Klondikers of global warming: men from all over the world who have come to Hammerfest, gateway to the Barents Sea, to make their fortune from new resources—oil, gas, fish and diamonds—made accessible by the receding ice" (Arctic booms 2005).

While the new Klondikers may come from all walks of life, recent scholarship—reinforced by personal observations made in fieldwork in Africa and South Asia—suggests to me that the harm that results from climate change and similar global phenomena will be disproportionately allocated to the weakest and poorest communities on the planet. I suspect that any benefits that accrue will tend to be disproportionately conferred on the rich and powerful.

Against this background, this chapter examines climate change from the perspective of human security—that is, freedom from fear and want at the level of the individual, family, or community. Human security is a concept that is often analyzed with reference to various measures of vulnerability. I begin the chapter by discussing the term *human security* as it was defined by the United Nations Development Program in 1994. I then argue that our networked world has introduced new forms of vulnerability and amplified some old forms (especially among certain sectors of humankind) that tend to undermine human security. I illustrate this with three case studies of the recent impact of climate change on communities in Bangladesh, Sudan, and the United States.[2] Finally, I conclude with the argument that improving human security by reducing vulnerability to harmful climate change and other transnational phenomena is as important as attacking the root causes of specific threats.

Self-interest alone might support this conclusion. No one wants to be the victim of a disease, crime, or act of terrorism that has its origins in a desperate and impoverished community whose land has been rendered uninhabitable by forces of environmental destruction that are beyond its control. But there is also a moral imperative to promote human security globally, at least for those who have linked their identities to environmentalism and hence to what Fritjof Capra (1997) terms the global "web of life."

As Mike Brklacich (2005), former director of the Global Environmental Change and Human Security Program, argues, it is unclear that the Framework Convention on Climate Change

(FCCC) and the Kyoto Protocol will succeed in halting or even slowing human-generated climate change. But even if we cannot stop trends like global warming in the near future, we can and should act to reduce the vulnerabilities of the world's least advantaged people, for these vulnerabilities reveal where climate change is first becoming a significant threat to safety and welfare.

Human Security

Scholars working to improve our ability to reduce economic and environmental vulnerability, including vulnerability to climate change, continue to debate how best to approach the problem. One approach is to think about human vulnerability to worsening economic and environmental conditions as a lack of "human security." Human security can be said to have two main aspects—safety from such chronic threats as hunger, disease, and repression and protection from sudden and hurtful disruptions in the patterns of daily life (UNDP 1994, 23). For Tariq Banuri (1996, 163–64),

> *security* denotes conditions which make people feel secure against want, deprivation, and violence; or the absence of conditions that produce insecurity, namely the threat of deprivation or violence. This brings two additional elements to the conventional connotation (referred to here as political security), namely human security and environmental security.

Some scholars are skeptical that human security can yield better insights than other approaches to the study of international development and security. Although the concept of human security has been criticized as being too broad to be analytically useful and has not had the immediate appeal of Robert Kaplan's "coming anarchy" thesis (1994), its development has been steady and has demonstrated considerable attraction for scholars, policymakers, and activists in the developing world and Europe.[3]

Large portions of humankind—primarily but not exclusively in the Southern Hemisphere—are rarely if ever free from danger and want. The concept of human security shines a spotlight on the economic and environmental factors contributing to human insecurity and the need, on moral, ethical, economic, and environmental grounds, to improve human security where it is most lacking. The fact that human security embodies a great deal may make it less analytically interesting to some scholars, but it would be wrong to suggest that there is not much analytical value in broad inclusive concepts that tell a compelling general story.[4] In many areas of academic inquiry, scholars aim to understand one part of a problem at a time, while others study the ways that the parts interact. In his analysis of the concept, Roland Paris (2001, 102) notes that a high level of inclusiveness can "hobble the concept of human security as a useful tool of analysis," but he ultimately concludes that "Definitional expansiveness and ambiguity are powerful attributes of human security. . . . Human security could provide a handy label for a broad category of research . . . that may also help to establish this brand of research as a central component of the security studies field."

Vulnerability is often used as a way of assessing human security and giving it greater analytical traction. Vulnerability can mean simply exposure to a hazard, but typically it involves some measure of adaptive capacity and other systemic properties. Vulnerability is thus commonly regarded as a function of exposure, sensitivity, adaptive capacity, and mitigation capacity and thus might be rendered as $Vf(E \times S \times A \times M)$. Much of this particular effort to focus the concept of human security and use it as a basis for analysis and policymaking has been undertaken by scholars and activists in the field of environmental security.[5] The arguments of the remaining pages follow their well-worn paths (e.g., Adger 1999; Khagram, Clark, and Raad 2003; Lonergan 1999; Matthew 2002b; Naumann 1996).

Networks and Vulnerability

Communication, commerce, and trade depend on networks of people, businesses, and governments. Understanding the characteristics of these networks helps scholars to identify trends in vulnerability to environmental stress.

The mathematician Albert-Lászlό Barabási (2003) argues that two types of networks exist. *Random networks* are composed of nodes that are linked to each other without pattern. In this type of network structure, each node tends to have about the same number of connections as all other nodes. The Internet originated as a random network with a small number of similarly interconnected nodes. The value of this type of network becomes limited as it grows because it quickly becomes cumbersome to navigate. As the ratio between noise and useful information increases, the transaction costs of using a random network rise dramatically.[6] Random networks tend to be used by a limited set of actors for specific purposes and are generally short-lived. For example, most people now use intermediaries—powerful search engines—to navigate the Internet and reduce electronic junk mail.

Far more common are what Barabasi (2003) calls *scale-free networks*, in which a relatively small number of nodes, referred to as *hubs*, develop with connections to a very large number of other nodes.[7] Scale-free networks are remarkably robust and resilient, they are notoriously difficult to regulate or neutralize, and they are easy to use or exploit. Their proliferation in recent years—moving people, ideas, beliefs, values, money, technology, and many other forms of capital among the world's 192 countries—has prompted a number of researchers on globalization to conclude that the power of the sovereign state has greatly diminished and that countries must now focus on managing the sometimes harmful local effects of powerful transnational networks that operate beyond their control.

Networks create enormous opportunities and benefits for large numbers of people, but they also introduce or amplify a specific type of vulnerability. For example, networks can be used to coordinate access to traditional forms of livelihood. Disruption of networks for such access can greatly reduce human security and reduce the ability of communities to adapt to changing economic and environmental conditions. Bad things can be transmitted quickly over vast distances, and individuals operating within a network are vulnerable to experiencing sudden and unexpected costs that may be very high. Scale-free networks intensify the formula $Vf(E \times S \times A \times M)$ in part by reducing the time available for adaptation and mitigation while multiplying the pathways for exposure. This is a process that many environmentalists have argued about for decades and is well captured in the familiar eco-adage, "Everything is connected to everything." The idea is that no one is ever completely immune to the potential negative social effects of transnational environmental change such as air pollution or ozone depletion. Exposure will not be even, and there will be geographic areas and time periods during which the risks are relatively high or low for one group or another. But often real harm cannot be predicted with enough precision to guide proactive policies or other responses. Indeed, at times harm may be inflicted on individuals and communities previously assessed to be at low risk. In a networked world, we are all vulnerable, all the time, although not at all equally.

Climate change fits this paradigm well and provides a good opportunity to develop it in two other directions. First, it is impossible to predict exactly when, where, and on whom climate change will inflict harm or good. From a local perspective, climate change might have a positive social impact as it brings rain to arid areas or warmth to cold areas and hence creates opportunities for agriculture and settlement. Some scenarios of climate change have generated considerable benefits for Canada and Russia, for example.

Second, climate change manifests itself as weather and thus has the potential to merge directly into transnational networks such as transportation systems. For example, a severe storm can snarl traffic, ground airplanes, and shut down ports quickly, leading to exorbitant economic costs as well as to death and injury.

The following three cases illustrate various ways in which local changes resulting from global climate change can threaten human security in one part of the world and affect other parts through social, economic, and environmental networks. Similarly, weakening of networks reduces access to critical resources, thereby increasing vulnerability to hunger and related woes.

Climate-Induced Flooding in Bangladesh

The conventional climate-change scenario for Bangladesh focuses on sea-level rise on its coastal region (figure 7.1). Bangladesh has "one of the most densely populated, low-lying, coastal zones in the world, with 20–25 million people living within a one-metre elevation from the high tide level" (Coastal

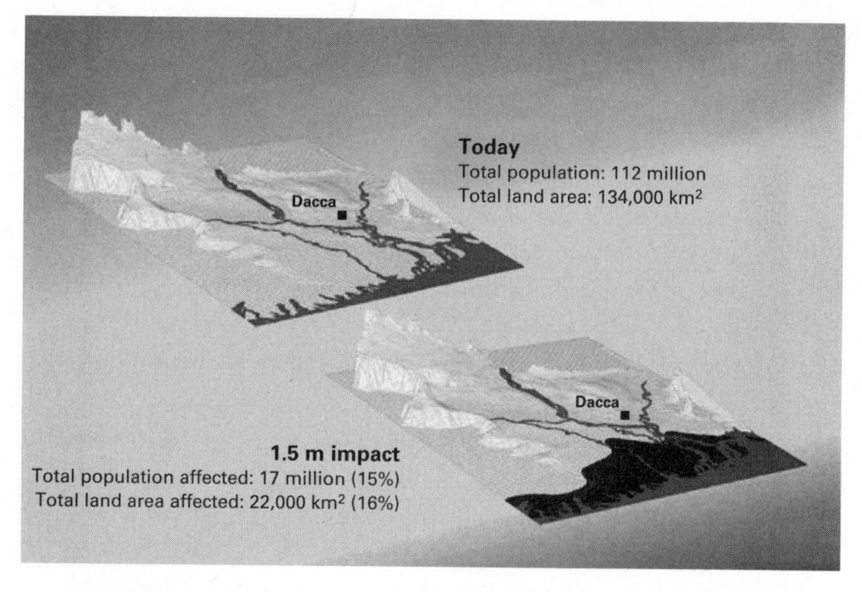

Figure 7.1
Potential impact of sea-level rise in Bangladesh
Source: UNEP/GRID (Geneva); University of Dacca; JRO (Munich); The World Bank World Resources Institute (Washington, DC).

Zone Management Studies n.d. 1–2). During the monsoon season of 2004, for example, the coastal region of Bangladesh was paralyzed by severe flooding. By mid-July, 60 percent of the country was under a blanket of water soiled with a rank mixture of industrial, agricultural, and household waste. Some 20 million people were directly affected, many of them facing food and freshwater shortages, skin infections, other diseases, and displacement.

Flooding on a similar scale had occurred sixteen years earlier, but at that time the cost of building a system of dams and dikes that could protect the country was deemed prohibitive by the government, so little was achieved to safeguard the population or its critical infrastructure. Combating climate change, the likely cause of the heavy snow melt in the Indian and Nepali Himalayas that contributed to the flooding, was equally out of reach (Huq 2004). The complex processes of global environmental change that amplify severe weather events, dense coastal populations, and widespread poverty have combined to make the people of Bangladesh highly vulnerable to the adverse effects of severe coastal flooding (World Bank 2000).[8]

But the vulnerability of the country is manifested each year in many different and less dramatic ways. For example, Tanguar Haor is a 9,727 hectare area of rich biodiversity that is protected as an internationally identified by treaty (or Ramsar) site but that also is home to 25,000 of the poorest people in the country.[9] All signatories to the 1971 Ramsar Convention are obliged to list at least one wetland as a "wetland of international importance" that will be protected in accordance with the Convention. The average per capita income in this region is $130 U.S.—about one third of the national average in Bangladesh and about 2 percent of the global average. Organized into forty-six villages that date back to at least the eighteenth century, the people are completely dependent on wetland resources—fish, reeds, forest products—for survival. The *haor* itself is formed of numerous ponds known as *beels*, which are inundated during the wet season when they merge into a single large lake. During this period, most of the inhabitants of the area migrate out, returning in the dry season to fish the *beels*, gather grass to create a variety of goods for themselves and for the market, and collect stones to sell for construction.

In the past sixty years, this area has changed its national status three times—from Britain to Pakistan to Bangladesh—but throughout this, de facto control has remained in the hands of the region's local elite. Most of the people living in the area have depended on informal, customary agreements for access to the *haor*'s resources. These agreements are far less respected today than they were in the past. The combined effects of colonial structures of inequality, rapid environmental change, dynamic population growth, a leasing system implemented in the 1970s that eroded customary rights and that most local people could not afford, and the state's decision to protect the area's biodiversity have all rendered the poor vulnerable to hunger, poverty, and violence.[10]

Increasingly fierce competition over access to the resources of the *haor* is affected by climate change that amplifies the wet and dry seasons, making *beels* more or less resource rich. Traditional livelihoods, insurance systems, and dispute-resolution mechanisms are undermined by these diminishing resources.[11] All too often the economic and political transformations that occur—partly in response to environmental changes—distribute costs to the poor and benefits to the rich. Facing abundance, for example, the rich find ways to channel resources and other goods into the market to make money, at the expense of the livelihoods of the poor. Faced with scarcity, the rich find ways to monopolize what is available, also at the expense of the poor.

Climate-Induced Drought in Sudan

Sudan has been independent since 1956 and has a population of 39 million. Like Bangladesh, it is one of the poorest countries in the world. While Bangladesh has spent a considerable amount of its gross domestic product on maintaining a large military because it fears aggression from India, Sudan has spent heavily on a twenty-one-year civil war that ended in

December 2004. As the civil war came to a close, however, regional violence in the northern state of Darfur escalated. As the economist Jeffrey Sachs (2005) writes,

> Failures of rainfall contribute not only to famines and chronic hunger, but also to the onset of violence when hungry people clash over scarce food and water. When violence erupts in water-starved regions such as Darfur, Sudan, political leaders tend to view the problems in narrow political terms. If they act at all, they mobilize peacekeepers, international sanctions and humanitarian aid. But Darfur, like Tigre, needs a development strategy to fight hunger and drought even more than it needs peacekeepers. Soldiers cannot keep peace among desperately hungry people.

Northern Darfur is one of three states in western Sudan. It has a population of about 1.5 million people who subsist primarily through farming and herding. Population growth has dramatically increased population density in the region, placing enormous pressure on its arid lands. In recent years, declining rainfall, likely a manifestation of global warming, has added considerably to the region's woes. Historically, the region was neglected by the more populous and oil-rich south, a neglect due partly to ethnic prejudice and partly to the civil war that has plagued the country for much of its existence. Ironically, as the civil war has diminished in recent years, culminating in the December 2004 peace agreement, years of drought have intensified long-standing conflicts over access to land and water between the farmers, who are also predominantly black and Muslim, and the herders, who tend to identify themselves as Arab Muslims.[12] The conflict has matured into a government-supported genocide perpetrated by Arab *janjaweed* (armed horsemen) against black farmers, who have antagonized the government by forming two dissident groups—the Sudan Liberation Army (SLA) and the Justice and Equality Movement (JEM). Survival strategies have included selling land, produce, and livestock cheaply so that these things will not be stolen and working the land unsustainably to extract nutritional value as quickly as possible. The region suffers from a severe food shortage. As the market serving the region has collapsed, poverty and malnutrition have intensified, and local residents have become highly vulnerable to diseases including malaria, yellow fever, cholera, and diarrhea. Since 2003, hundreds of thousands of people have died or been displaced.

In *Hegemony or Survival*, Noam Chomsky (2003, 2) writes that "humans have demonstrated that [destructive] capacity throughout their history, dramatically in the past few hundred years, with an assault on the environment that sustains life, on the diversity of more complex organisms, and with cold and calculated savagery, on each other as well." Sudan is a harsh case in point.

As in many parts of the world, conflict in Sudan is linked strongly to economic and ethnic factors operating at the local level. But corrupt government, ethnic rivalry, intransigent poverty, and an arid environment also create great vulnerability to the transnational problem of climate change, which enters into Sudan as drought and amplifies both the vulnerability of the people and the multiple forms of violent conflict and misery to which they are more or less permanently exposed. In Sudan, as in Bangladesh, the sustaining elements of vulnerability need to be addressed so that the people can adapt to or act to mitigate any adverse effects of climate change. For example, research is needed to identify how existing scale-free networks connected to Sudan can create benefits across social and political divisions through reducing the economic and environmental insecurities of the poor.[13]

Figure 7.2
History of movement of Hurricane Katrina
Source: ⟨http://www.accuweather.com⟩. © 2005 AccuWeather, Inc.

Hurricane Katrina

Much of the city of New Orleans lies below sea level in southeast Louisiana. Surrounded by the Mississippi River and Lakes Pontchartrain and Borgne, the city has been vulnerable to flooding since its founding in 1718, and an extensive system of levees and canals was constructed to provide protection. When Hurricane Katrina hit the city on August 29, 2005, its levees were breached, and within twenty-four hours 75 percent of the city was under water (figure 7.2). Analysts have assembled a complicated story of how factors such as the degradation of the Mississippi River delta and extensive ground subsidence (caused in part by the canals designed to drain the city) combined to enhance New Orleans' vulnerability to a sustained category 3 storm. Equally important is the continuing investigation into the failures of the city's advance-warning, disaster-response, and clean-up operations.

Although 70 to 80 percent of the residents evacuated the New Orleans area, relying largely on private transportation, tens of thousands did not. Those remaining in the area during the storm were predominantly poor or elderly, and most were African Americans. In the disaster zone, response teams were unable to communicate effectively, evacuation efforts were confused, many citizens were stranded, others were packed into the squalid and threatening conditions of the Superdome, and by March 2006 the confirmed death toll had reached 1,293. Approximately 1 million people in the Greater New Orleans area required emergency services; some 1.3 million were displaced by the flooding. The main employers of the city—oil production, gambling and entertainment, and agriculture and forestry—were severely disrupted by the hurricane.

The extensive harmful effects of this natural disaster, which may endure for decades, have weighed especially heavily on the city's poor. As Naomi Klein (2005) writes, "Wealth in New Orleans buys altitude. That means that the driest areas are the whitest (the French Quarter is 90 percent white; the Garden District, 89 percent; Audubon, 86 percent; neighboring Jefferson Parish, where people were also allowed to return, 65 percent)." Not only did the flooding affect neighborhoods inhabited largely by poor people, the elderly, and African Americans, but these are also the people who are finding it most difficult to return to their homes.

The principal lessons of Hurricane Katrina are as applicable to Sudan and Bangladesh as they are to New Orleans. While no single event can be attributed to climate change per se, severe drought and weather events are consistent with climate-change models (Knutson and Tuleya 2004). Such phenomena can affect wealthy countries as well as poor ones, but in both cases it is likely to be subcommunities of the poor and elderly who suffer the most and recover the most slowly.

Conclusions

Policymakers like simple cause-and-effect relationships and concrete problems. It is easier to identify and address a tangible threat such as HIV-AIDS, SARS, or Asian bird influenza than it is to improve public health or manage the problem of infectious disease generally. The complex, transnational networks that have come to shape the lives of people around the world further complicate the difficulty of addressing vulnerability rather than threat. Vulnerability is now deeply implicated in structures that are agile and resilient, that defy regulation, and that are readily accessed—by employers, money launderers, and amoebic parasites. Although networks can provide opportunities and amplify vulnerabilities, the poor have tenuous and weakening connections to the networks that grant opportunities.

Today, approximately one fifth of humankind survives on 1 percent of the world's wealth, while another fifth lives on some 80 percent of it. Enormous gains have been made in areas such as infant and child mortality, life expectancy, and literacy—the variables that are used to measure human security and human development. But equally enormous asymmetries of power and wealth have continued to exist. The combined wealth of the richest ten people on the planet is greater than that of the poorest 20 percent of the world's population (1.3 billion people). In many ways, we are more connected to each other than ever before, but our economic circumstances vary dramatically, and we are not equally vulnerable to transnational forces such as climate change.

In this context, it is easy for the rich to focus on narrowly defined threats and leave the solutions to the far more elusive issue of vulnerability to future generations. The plight of the poor—in Bangladesh, Sudan, or southeastern United States—as they are driven from their homes and jobs by floods or droughts that have been amplified by climate change may be unsettling to watch on CNN, but the television set can be turned off. Nevertheless, the interconnectedness of the world compels us to ask, How vulnerable might we be to climate change or to its adverse social effects? Where will desperately poor people—who are unemployed, landless, and locked into bloody ethnic struggles—turn for help? And what moral obligation do we have when a process of global change in which we are deeply implicated places great burdens on people who have had a fairly negligible impact on the global change itself?

There are no simple solutions to climate change or to the suffering in Sudan or the misery in Bangladesh. But California and the Netherlands have taken concrete actions to reduce their vulnerability to climate change, which demonstrates that some positive preventive steps can be taken. While linking what is possible to what is needed is never easy, failing to do so may be unsustainable from both interest-based and moral perspectives.

Notes

1. For discussion, see Matthew and Shambaugh (1998).

2. Another good example is Vietnam (Adger 1999).

3. See, for example, Thomas and Wilkins (1999), Tehranian (1999), Suhrke (1999), and Yuen (2001). A more explicit union of environmental security and human security is evident in Nauman (1996).

4. Concepts such as class relations, human rights, and democracy are broad and inclusive and do an enormous amount of work in contemporary political analysis.

5. Details available at ⟨http://.gechs.org⟩ (accessed April 9, 2007).

6. Imagine, for example, if all airports had about the same number of connections. They would tend to be small, at least compared with the huge airports of major cities. Negotiating such a network of airports could be painfully slow: travel from Los Angeles to Geneva might require a dozen stops along the way as one hopped from one small airport to the next.

7. In a scale-free network of airports, many small airports are connected to a few large airports that will have many connections. A trip from Los Angeles to Geneva might require only one or two stops.

8. The Bangladesh experience contrasts with that of other low-lying regions, such as the Netherlands, which Bangladesh turned to for assistance in the 1990s. Unfortunately, it was not able to afford the infrastructure the Dutch have successfully constructed to protect their land. Similarly, the record rainfall in arid southern California in 2004 and 2005 caused little harm because over $200 million had been invested in flood-control infrastructure following the heavy rains of 1997 and 1998.

9. This discussion of Tanguar Haor is based on field research conducted in Bangladesh as part of the International Union for Conservation of Nature and Natural Resources (IUCN) Sustainable Livelihoods, Environmental Security and Conflict Mitigation in South Asia. For an overview of this study, see Matthew (2005).

10. On ecological collapse, see Diamond (1994).

11. A traditional insurance system in the Haor is to engage the family in multiple activities such as fishing, stone collection, grass collection, and so on. The idea is that to specialize in one area (e.g., grass collection or fishing) is too risky because if that endeavor fails, then the family faces a catastrophe.

12. For an influential quantitative study linking environmental change to other factors to better explain and predict violent conflict, see Collier (2000).

13. This section is based on research conducted for the International Institute for Sustainable Development (IISD) project on Climate Change, Resources and Conflict: Understanding the Links between

Environment and Security in Sudan. The author is a member of the project team. Research information can be accessed at ⟨http://.iisd.org/natres/security⟩.

References

Adger, Neil. 1999. Social vulnerability to climatic change and extremes in coastal Viet Nam. *World Development* 27: 249–69.

Arctic booms as climate change melts polar caps. *Observer*, November 27. ⟨observer.guardian.co.uk/international/story/0,6903,1651724,00.html⟩ (accessed April 26, 2006).

Banuri, Tariq. 1996. Human security. In Naqvi Nauman, ed., *Rethinking security, rethinking development*, 163–64. Islamabad: Sustainable Development Policy Institute.

Barabasi, Albert-Laszlo. 2003. *Linked: The new science of networks*. New York: Plume.

Brklacich, Mike. 2005. Climate change and human security. Paper presented at the University of California, Irvine. February 15.

Capra, Fritjof. 1997. *The web of life*. New York: Anchor Books.

Chomsky, Noam. 2003. *Hegemony or survival*. New York: Henry Holt.

Coastal Zone Management Studies. n.d. Bangladesh. Coastal Zone Management Studies. ⟨http://ccasia.teri.res.in/country/bang/impacts/impacts.htm⟩ (accessed March 11, 2005).

Collier, Paul. 2000. Economic causes of civil conflict and their implications for policy. A paper prepared for the World Bank Group. ⟨http://.worldbank.org/research/conflict/papers/civilconflict.pdf⟩ (accessed April 7, 2007).

Diamond, Jared. 1994. Ecological collapse of past civilizations. *Proceedings of the American Philosophical Society* 138: 363–70.

Huq, Saleemel. 2004. Bangladesh floods: Rich nations must share the blame. Science and Development Network, London. ⟨http://.scidev.net/Editorials/index.cfm?fuseaction=readEditorials&itemid=125&language=1⟩ (accessed March 11, 2005).

Kaplan, Robert. 1994. The coming anarchy: How scarcity, crime, overpopulation, tribalism, and disease are rapidly destroying the social fabric of our planet." *The Atlantic Online.* ⟨theatlantic.com/politics/foreign/anarchy.htm⟩ (accessed March 11, 2005).

Khagram, Sanjeev, William C. Clark, and Dana Firas Raad. 2003. From the environment and human security to sustainable security and development. *Journal of Human Development* 4(2): 289–313.

Klein, Naomi. 2005. Purging the poor. *The Nation.* ⟨http://.thenation.com/doc/20051010/klein⟩ (accessed March 11, 2005).

Knutson, Thomas R., and Robert E. Tuleya. 2004. Impact of CO_2-induced warming on simulated hurricane intensity and precipitation: Sensitivity of the choice of climate model and convective parameterization. *Journal of Climate* 17(18): 3477–95.

Lonergan, Steve. 1999. Global environmental change and human security science plan. IHDP Report 11. Bonn: IHDP.

Matthew, Richard A. 2002b. In defense of environment and security research. *Environmental Change and Security Project Report* 8 (Summer): 109–24.

Matthew, Richard. 2005. Sustainable livelihoods, environmental security, and conflict mitigation: Four cases in South Asia. *IUCN Poverty, Equity and Rights in Conservation Working Paper Series*, Gland, Switzerland. ⟨http://iucn.org/themes/spg/Files/IUED/Case%20Study%20South%20Asia.pdf⟩ (accessed March 11, 2005).

Matthew, Richard, and George Shambaugh. 1998. Sex, drugs and heavy metal: Transnational threats and national vulnerabilities. *Security Dialogue* 29 (Summer): 163–75.

Nauman, Naqvi, ed. 1996. *Rethinking security, rethinking development.* Islamabad: Sustainable Development Policy Institute.

Paris, Roland. 2001. Human security: Paradigm shift or hot air? *International Security* 26 (Fall): 87–102.

Sachs, Jeffrey. 2005. Climate change and war. *TomPaine.com*, Washington, DC. ⟨http://.tompaine.com/print/climate_change_and_war⟩ (accessed March 11, 2005).

Suhrke, Astri. 1999. Human security and the interests of states. *Security Dialogue* 30 (September): 265–76.

Tehranian, Majid, ed. 1999. *Worlds apart: Human security and global governance.* London: Tauris.

Thomas, Caroline, and Peter Wilkins, eds. 1999. *Globalization, human security and the African experience.* Boulder, CO: Lynne Reinner.

United Nations Development Program (UNDP). 1994. *Human Development Report 1994*. Oxford: Oxford University Press.

World Bank. 2000. Bangladesh: Climate change and sustainable development. Report No. 21104 BD. 〈wbln1018.worldbank.org/sar/sa.nsf/Attachments/cvr/$File/cvr.pdf〉 (accessed March 11, 2005).

Yuen, Foong Khong. 2001. Human security: A shotgun approach to alleviating human misery? *Global Governance* 7 (July/September): 231–36.

8

Climate Change: What It Means to Us, Our Children, and Our Grandchildren

Joseph F. C. DiMento and Pamela Doughman

Climate change is a complex challenge, perhaps one of the largest that the world has ever faced. This book has presented some important findings about climate-change science and related social and policy issues.

We have described changes in climate, including those caused by human activities, and explained why there is a gap between scientific understanding of climate and actions taken by society to remedy the harmful effects of emissions-related climate changes. We have described why climate change has been difficult to understand, based on the characteristics of science, on communication difficulties, and on political motivations.

In the first chapters, we described the dynamics of climate and the greenhouse effect, presenting a primer on the science and an historical overview of scientific discoveries. We distinguished between climate and weather and described the components—from clouds to oceans—and interactions that compose the climate system and human contributions to greenhouse gases. We noted *how* scientists know and introduce some of the people who have discovered what humans can do to affect the climate—over centuries and even years. The earth has become warmer and may grow warmer without major changes in the amounts and kinds of energy sources we use and how we use them.

Climate change poses challenges for local, national, and international governments and the private sector. Further changes in average temperature, precipitation, and weather events will affect human health and global and regional economies. Ecosystems will change, and some species will be made extinct. Ice sheets will melt, glaciers will disappear, and oceans will continue to rise.

Climate change will be different in different regions. Agriculture will be more productive in some zones and jeopardized in others. Some areas, like the Arctic, will continue to face dramatic alterations that threaten lifestyles and the viability of flora and fauna. Within the United States, changes will be felt in different ways in the Pacific Northwest, California, the Midwest, the East, and along the many thousands of miles of coast. Changes in weather, water availability, crop yields, heat waves, and public health—and the associated demands for energy—will be significant in some places at even low temperature increases. At the high level of 5.8°C (10.5°F), the impacts will be severe.

In chapter 4, we offered an historical perspective on what scientists know. The impressive research record comes from the work of individual laboratories and the compilations and assessments of existing knowledge by the Intergovernmental Panel on Climate Change (IPCC) and other major scientific bodies. Even so, there continues to be a mistaken view among some of the public that there is no consensus on whether climate change exists and that some scientists have self-interested reasons for viewing the situation with alarm. The future of climate change is a phenomenon that is not knowable with certainty, but current and probable trends are risky to ignore.

Scientists do some things very well, but communication of their work to the public is not high on the list of their skills—or desires. Nonetheless, as is shown in chapter 4, whatever some people believe and contrarians assert, the reality of climate change and its effects stands up well to the generally accepted standards of scientific inquiry. Denying that global warming is real is simply a refusal to look at the evidence.

Many individuals, private organizations, states, and national and international groups consider the record sufficiently clear to act, even though some think that more research and analysis are needed before major shifts are made. The latter are driven in part by their assessments of the costs associated with actions to control emissions. In chapter 5, the current and future actions of international efforts such as the Framework Convention on Climate Change and the Kyoto Protocol are examined.

Some countries have gone beyond the minimum requirements of the Kyoto Protocol and have demanded energy efficiency, renewable energy, new products for business and the consumer, and markets for the trading of emissions credits. Others have exited the process, either explicitly or by ignoring the need for controls until more research is done. Some governments have used legal means (legislation and lawsuits) to force other governments and businesses to decrease their emissions of greenhouse gases. Business has been involved, sometimes by choice, in recognizing the potential sources of profits in emissions-reducing products and processes. The long periods of time that greenhouse gases remain in the atmosphere mean that some level of climate change is already unavoidable, but harmful effects can be mitigated by taking actions—ranging from sustainable forest management to energy-saving residential and commercial building practices.

Communicating information about science is the mission of the Newkirk Center for Science and Society, which sponsored publication of this book. All of the chapters address the difficulties in translating climate-change science to nonscientists and the attentive public, but chapter 6 makes these challenges explicit by comparing them to other topics—including other environmental topics—that clamor for public attention. Some of the major impediments to effective climate coverage lie in the newsroom and in the nature of news itself. Climate scientists are not describing something that has front-page impact. Journalists—who are driven by time, space, and format pressures—at times can confuse readers into believing that human contributions to changes in climate are a matter of great scientific debate. They are not.

The localized effects of climate change will vary. Not everyone will be hurt, and some will likely benefit. But as chapter 7 notes, some regions are extremely vulnerable to ongoing changes in climate, and their populations, their resources, and their environments will be damaged in significant ways. Effects in the truly poorest communities may well be felt worldwide through the economic networks that have evolved in a globalized world.

Changing Knowledge

Our understandings of the challenges, risks, and opportunities of climate change evolve daily as new scientific information becomes available for society to consider. Abrupt climate change has moved from science fiction to a subject of serious attention. One recent analysis, for example, concludes we have perhaps ten to thirty years to keep carbon levels from doubling and avoid rapid, extreme climate change (Baer and Athanasiou

2005.) The links between climate change and major climate and weather events are better understood than they were a few years ago. Waters west of Africa that are a "nursery for tropical storms" have warmed over the last hundred years in a way that appears largely to be a result of human-influenced global warming. Increases in temperature of a few tenths of a degree could increase hurricane strength as those storms draw energy from tropical waters. Scientists at the Commerce Department's Geophysical Fluid Dynamics Laboratory found that computer simulations that took into account emissions of heat-trapping gases had "replicated the warming trend much more realistically than did simulations that used only natural fluctuations in the climate" (Revkin 2006, 18).

Some information is not new scientifically but nonetheless important because it makes graphic, local, and specific the general effects that were reported earlier: "While officials argue over carbon emission controls and global warming treaties, tree farmers, . . . gardeners, anglers and bird-watchers sense the change in the air" (Hotz 2006, 1). Spring is coming three weeks earlier in parts of North America, and springs and winters are becoming milder. From 1950 to 1993, the coldest winter temperatures rose by 2.8°C (5°F), the warmest spring temperatures rose 1.8°C (2.5°F), and average temperatures in New England cities increased 1.1° to 1.9°C (2.0° to 3.4°F) (Hotz 2006, 1). Cherry trees flower on average two weeks earlier than they did a generation ago, rivers and streams reach their high-water levels as much as ten days earlier than fifty years ago, ducks return to Massachusetts a month earlier than three decades ago, hummingbirds arrive eighteen days sooner, and sugar maples are tapped weeks earlier than a few decades ago (Hotz 2006, 1).

Results from over eight hundred scientific papers on ecology offer a virtually uniform picture of the changes in species vitality and distribution: "People are finally starting to see the changes, spread across the world from the tropics to the Arctic and across every taxonomic group," notes ecologist Camille Parmesan of the University of Texas at Austin, who reviewed the behavior of seventeen hundred species of plants and animals (Hotz 2006, 1).

At one time, adapting to climate change—as opposed to acting now to decrease greenhouse-gas emissions—was almost an ideological position. Refusal to discuss adapting to climate change was presented by some as denying scientific facts or avoiding problems because they were not convincingly presented or might cost too much and produce too little. Now experts in the United Kingdom, whose government has taken the lead in innovative climate policy, conclude: "An equitable international response to climate change must include action on both adaptation and mitigation. Adaptation and mitigation are not choices: substantial climate change is already inevitable over the next 30 years, so some adaptation is essential" (Stern Review Team 2006, 4).

But there is much more to know about the effects of climate change and about the costs associated with various responses to it. Should businesses begin emission reductions now in a voluntary way? Should governments make costly changes without knowing how other governments will respond? What good do individual consumer decisions accomplish in a world that has billions of sources of greenhouse gases? For example, in thinking through the wisdom of entering into carbon markets, there are questions about the price per ton of carbon that businesses include in their assessments. Uncertainty comes in part because of lack of information but also for fundamental reasons related to governance in a period of changing views of rights: "Certain types of rights, such as rights to emit greenhouse gases, . . . could be delineated without great difficulty . . . but [o]ther rights, such as credits for carbon stores in the soil and trees of a forest stand or in the ocean, would be more complicated to define" (Congressional Budget Office 2003).

Furthermore, unless major disruptions are clearly linked to climate change, the issue may fade as nations and international organizations address other more pressing problems (such as SARS, avian flu, and terrorism). Even scientists may move to other areas of interest. The National Academy report noted in chapter 3 worried that the best scientists might lose interest in participating in the IPCC process, which would mean that a central vehicle for processing and communicating scientific information about their climate research would be jeopardized.

The Economics of Climate Change and Risk Assessment

Through its nongovernmental organizations, businesses, states, and international communities, society is learning more and more about climate change. It needs to engage in decision making under uncertain conditions about atmospheric dynamics and causes and effects. How governments will actually respond to evolving information is an ongoing question, although governments are increasingly saying they will act. (The difference between declarations and actions is most disturbingly reflected in the gaps between the Kyoto Protocol emission-reduction targets and actual performance by nations in the past decade.) Furthermore, we do not fully understand the actual costs and benefits of meaningful responses to climate change.

The risks associated with climate change are assessed differently by people who vary in their assessment of, tolerance for, and acceptance of risks. Numerous, noisy, and sometimes countervailing conclusions are reached by entities that influence public understanding. The media wax hot and cold about environmental issues, although the trend recently has been a strong set of communications that emphasize the hot. *Time* magazine, *National Geographic*, *Rolling Stone*, *60 Minutes*, the *New York Times*, the *San Francisco Chronicle*, the *Los Angles Times*, the *New Yorker* have all published major articles on what they characterize as the serious nature of climate change. Nonetheless, the media seem structurally required to keep the issue as an ongoing debate. A few vocal and influential scientists remain contrarians and are given disproportionate attention in the media, in legislative committees, and in other policy forums. Business increasingly talks green, but consumers sometimes experience only a greenwash. Governments differ in how they state the problem or whether they think there is one. UK Prime Minister Tony Blair and California Governor Arnold Schwarzenegger speak with a sense of urgency, whereas U.S. President George W. Bush has described a commitment to absolute emission reduction as more risky than climate change.

There are different risks for different regions. The weakest and poorest communities will feel the harm most. But here there is another complicating factor as citizens learn about the possible spillover of serious effects across regions—from migration, from movement of environmental health problems, and from the political unrest that is linked to serious environmental degradation.

Environment and Economics

Understanding the environmental effects of climate change is necessary but not sufficient. We must also understand the effects of climate change on the economy and vice versa.

The Function and Complexity of Cost-Benefit Analysis

For decades, economists have tried to assist policymakers in decisions by conducting cost-benefit analyses of proposed actions. There is no exception in the area of climate change. The Congressional Budget Office (2003) reports that "Over the past 15 years, a large number of studies have analyzed the potential costs and benefits of averting climate change. Some researchers have attempted to incorporate the studies' results in global and regional models of economic growth and climate effects and have used models to conduct so-called integrated assessment of policy proposals related to climate change. They have also estimated the cost of emission control policies that would yield the greatest net benefits in terms of economic growth, reduced emissions, and the resulting climate effects."

Rather than showing a range of possible conditions, some studies make a number of controversial assumptions: "technological change will probably lower the cost of controlling emissions," "people are likely to be wealthier in the future," "carbon ... emissions that occur sooner rather than later will have more time to be absorbed from the atmosphere by the oceans." These are assertions worth debating and help create a context for making decisions on how to proceed.

Costs and benefits are also influenced by what economists call a *discount rate*. A dollar today is worth more than a dollar tomorrow; people place less value on the future than they do on the present. As the Congressional Budget Office (2003) points out, "at discount rates that approximate market rates, even very large long-term costs and benefits are dramatically devalued.... The choice of discount rate therefore makes a huge difference in thinking about long-term problems such as climate change."

The central question for individual countries is reflected in actions that are taken at intergovernmental meetings: how the costs and benefits of climate change will be distributed: "Policies that balance overall costs and benefits do not necessarily balance them for every person, and policies that maximize the net benefits to society do not necessarily provide benefits to each individual. A policy may yield positive net benefit by causing both very large aggregate losses and only slightly larger aggregate gains" (Congressional Budget Office 2003). There is distribution regionally and distribution across generations. The international conferences of the parties (COPs) on climate change address this issue explicitly often and implicitly always. Debate continues over the roles that are played by India, China, and other rapidly developing large countries in climate change. Although greenhouse-gas emissions in these regions were relatively low in the past, they are expected to grow quickly in coming decades unless carbon-reducing strategies and technologies are deployed quickly.

Integrated assessments can also help, although they sometimes make controversial assumptions. A study by William Nordhaus (2001) is an example of how this technique can be used to analyze tradeoffs. Based in part on a large literature review, "The study concluded that modest warming of up to 2.3°F (1.3°C) would have essentially no net impact on the world economy and might even yield some net benefits. But the study also concluded that in the absence of efforts to reduce emissions, the average global temperature would rise by about 3.6°F (2.0°C) over the next century and by 6.1°F (3.4°C) over the next two centuries. These changes would inflict damages—measured as a reduction in world economic output—of roughly 1.0 percent (about $1 trillion in 2000 dollars) in 2100 and about 3.4 percent (nearly $7 trillion) in 2200" (Congressional Budget Office 2003). Losses were quantified for the areas of impact summarized throughout this book. The study was highly sensitive to assumptions made and "its weighing of the welfare of current generations against that of future generations" (Congressional Budget Office 2003). To make the study's results more graspable, it noted, for example, that "an emissions charge of $10 per metric ton of carbon would add . . . about 2.5 cents to the price of a gallon of gasoline."

Some of the economic questions on cost arise from speculations and best guesses about how governments and the private sector will implement programs to address climate change. Some programs have begun, and some have faltered. Greenhouse-gas markets have been begun in the European Union and Japan, and in Chicago a voluntary market is in place. The Kyoto Protocol's clean-development mechanism, which allows developed countries to earn credits by reducing greenhouse-gas emissions in developing countries, has made some progress so far with projects in the developing world, but adoption of actual successful projects has been slow. In the United States, East Coast states have made a commitment to a cap and trade program, and California has proposed a greenhouse-gas performance standard. Some states have sued utilities to characterize these emissions as a public nuisance, and other states have brought suit to force the U.S. Environmental Protection Agency to regulate greenhouse-gas emissions from coal-fired plants. (The EPA has argued that it does not have the authority under the federal Clean Air Act to regulate carbon dioxide and that because the gas does not directly affect human health, like already regulated pollutants, it cannot be classified as a pollutant.) If lawsuits like these are successful, the playing field for business decisions will be dramatically changed to substantially reduce climate emissions.

Some costs are getting clearer as we experience (rather than simply wonder about) climate change. But the costs of climate change remain a matter of guesses, estimates, assumptions, and—to a certain extent—ideology. We know where climate-change costs are likely to be felt most acutely, but predicted impacts at local or regional levels are by necessity very general. In agriculture, there will be increased costs (and, in places, benefits) associated with shifts in the type of crop cultivated and the water available for those crops. Industry, whether carbon dioxide is regulated directly or not, will need to pay for ground-level ozone noncompliance—through stricter pollution controls or fines. A less healthy workforce (with increased asthma and other respiratory and heat-related illnesses) will take more sick days, decrease business productivity, and require higher health premiums. Energy will be more costly, and peak demand will be higher in some regions. Forest fires will be more frequent and more severe.

The estimates of the costs of reducing greenhouse gases vary with time scale and with understandings of the value we put on things that will happen to us or to our children and grandchildren in the future. The Congressional Budget Office summarized the situation blandly but accurately: "Climate policy thus involves balancing investments that may yield future climate-related benefits against other, non-climate related investments—such as education, the development of new technologies, and increases in the stock of physical capital—that are also beneficial. If climate change turned out to be relatively benign, a policy that restricted emissions at very high expense might divert funds from other investments that could have yielded higher returns. Conversely, if climate change proved to be a very serious problem, the same policy could yield a much higher return" (Congressional Budget Office 2003).

Estimates will be influenced by changes in the economy that may include the costs and benefits of renewables and efficiency. At some point, consumers will either see and value savings associated with alternative energy sources and consumer products (hybrids, public transport, the hydrogen highway, and biofuels) or be disappointed that the predictions of environmentalists or the fears of economists (gas at $5 per gallon, for example) did not come to pass.

The economics of climate change can work for society in more direct ways, including through the many opportunities for the business sector. For example, the California Climate Action Team report to the governor recommended a suite of strategies to increase investments in low-carbon technologies—such as energy efficiency, electricity and transportation fuels from renewable resources, diesel anti-idling, recycling 50 percent of waste statewide, and smart farming, land-use, and transportation planning. The strategies include voluntary actions, incentives, and regulatory programs.

An economic assessment by the California Climate Change Center (2006) looked at eight policies—building efficiency, vehicle emissions standards, HFC reduction, manure management, semiconductors, landfill management, afforestation, and cement manufacturing—that would achieve half of the governor's goal of reducing greenhouse-gas emissions to 1990 levels by 2020. The report concluded that "the aggregate economic benefits of many GHG mitigation policies outweigh their microeconomic costs, and more generally that it is possible for the state to reconcile growth and environmental objectives.... Many policies under active consideration... actually *save money* and *increase employment* overall because the indirect effects are so important. These overall benefits only become apparent when the economy-wide implications... are taken into account" (California Climate Change Center 2006, 10–16). The report said that the economic benefit from these eight actions would be $60 billion (2.4 percent of gross state product) and would add twenty thousand jobs by 2020.

Investment firms and funds can focus on companies that produce technologies and services that help mitigate or adapt to climate change. The California State Teachers Retirement System (CalSTRS), with more than $130 billion in assets, is working to improve corporations' climate-risk disclosure and response as part of its work on improving performance of its investment portfolio. CalSTRS was one of twenty large public and private U.S. investors (collectively controlling more than $800 billion in assets) that pressured thirty large publicly held insurance companies to disclose financial risks of climate change for life, health, and property insurance profitability. The investors also asked the insurance companies to identify steps they are taking to reduce exposure to these risks and explore new business opportunities in the changing economic environment expected from climate change. The investors requested climate-risk reports by August 2006 (Ceres Group 2005). Insurance agencies that take climate into consideration also can develop new types of policies and cancel less profitable coverage.

But risks and opportunities are not limited to insurance. Many businesses are addressing risks and identifying opportunities related to climate-change mitigation. Businesses that have joined the California Climate Change Registry or an EPA partnership as early actors may be saving more money than the laggards. Recognizing that business can respond to new markets, some of the climate-change programs aim to "make environmental protection look more like an ordinary business issue to managers and can allow them to apply risk management tools, perhaps in the form of financial derivatives." The Berkeley report (California Climate Change Center 2006) characterized a cap-and-trade program as one such example. In a *cap-and-trade program* (also known as *emissions trading*), businesses have a limit to the greenhouse gases that they can emit but are allowed to buy credits from other businesses if they exceed their limit or sell their credits if they emit less than their cap. Businesses are free to determine how they want to meet their cap. Even without a cap-and-trade system, mitigation is already a business opportunity that could have significant profit potential. If the overall approach to climate continues to favor developing countries and not demand their direct contributions to emissions reductions, devices such as the clean-development mechanism could become a considerable source of income for developing countries.

The Issue in Context

What climate change means for you, your children, and your grandchildren depends on which generation you are considering, on where you live, and on how you value society's goods and bads. But as we have seen, it also depends on the actions that are taken by the international organizations, nations, states and municipalities, regions and provinces, and businesses that directly influence climate policy. Citizens can also play a role in helping to mitigate and respond to climate change at home, in business, and through participation in government processes. Policymakers increasingly include climate change in their view of the world. Climate change is becoming part of the calculus on whether to switch to a low-carbon automobile fleet or personal car, to insure or refuse to insure, to invest in renewables or exploration of nonrenewables, to minimize the environmental footprint of new homes or not: the list is almost as long as the sources of greenhouse gases.

References

Baer, Paul, and Tom Athanasiou. 2005. Honesty about dangerous climate change. ⟨http//www.ecoequity.org/ceo/ceo_8_2.htm⟩ (accessed April 7, 2007).

California Climate Change Center. 2006. Managing greenhouse gas emissions in California. California Climate Change Center, University of California, Berkeley. ⟨http://calclimate.berkeley.edu⟩ (accessed December 6, 2006).

California Environmental Protection Agency, Climate Action Team. 2006. *Climate Action Team Report to Governor Schwarzenegger and the California Legislature.* ⟨http://www.climatechange.ca.gov/climate_action_team/reports/2006-04-03_FINAL_CAT_REPORT.PDF⟩ (accessed April 7, 2007).

Ceres Group, Inc. 2005. Investors turn up heat on insurers to disclose risk from climate change. *Insurance Journal*, December 1. ⟨http://www.insurancejournal.com/news/national/2005/12/01/62669.htm⟩ (accessed December 6, 2006).

Congressional Budget Office. 2003. The economics of climate change. Congressional Budget Office, Washington, DC. ⟨http://www.cbo.gov/showdoc.cfm?index=4171&sequence=3⟩ (accessed December 6, 2006).

Hotz, Robert Lee. 2006. Tapping into a changing climate. *Los Angeles Times*, April 32, A1.

Nordhaus, William D. 2001. Economics and policy issues in climate change: The economics of the Kyoto-Bonn accord. *Science* 294(5545): 1283–84.

Revkin, Andrew C. 2006. Nursery for hurricanes. *New York Times*, May 2, A18.

Stern Review Team. 2006. What is the economics of climate change? Discussion Paper 31 (January). Report to the Chancellor of the Exchequer and the Prime Minister, London. ⟨http://www.hm-treasury.gov.uk/media/213/42/What_is_the_Economics_of_Climate_Change.pdf⟩ (accessed June 22, 2006).

Glossary

abrupt climate change A large, rapid, unexpected change in average weather conditions that affects local, regional, and global patterns.

acid rain or precipitation The removal of acidic gases and ***aerosols*** from the atmosphere to the earth's surface by fog, dew, rain, and snow. The main source of acid-forming gases is the release of sulfur dioxide and nitrogen from the burning of fossil fuels.

aerosols Suspended solid and liquid particles that range in size from tiny molecular clusters to particles visible to the human eye. Some are man-made, spewed by smokestacks, unfiltered tailpipes, and volcanoes. They can have both cooling and warming influences on the ***atmosphere***.

albedo The reflectivity of a material. According to the ***Intergovernmental Panel on Climate Change (IPCC)***, albedo is the fraction of solar radiation reflected by the surface or object, often described as a percentage.

anthropogenic Human-caused or -related.

atmosphere The envelope of gases that surrounds the earth or other planets.

atmospheric circulation Atmospheric motion that is a response to the unequal heating of the earth's surface; redistribution that acts to create energy balance.

biosphere The part of the earth system that includes all ***ecosystems*** and living things on land, in the ***atmosphere***, and in the oceans.

cap and trade A program in which an entity, such as a business, has a limit on a substance, such as ***greenhouse gases***, that it can emit but is allowed to buy credits from other entities if it exceeds its limits or sell its credits if it emits less than its cap.

carbon An abundant chemical element essential for life. In the form of ***carbon dioxide*** *(CO_2)* in the atmosphere, it acts as a ***greenhouse gas***, raising the temperature of the earth. See also ***carbon dioxide***.

carbon cycle The exchange of ***carbon*** among the atmosphere, oceans, and land, including the ***biosphere*** and ***lithosphere***.

carbon dioxide (CO_2) A heavy, colorless, nonflammable gas that makes up about 0.04 percent of the earth's ***atmosphere***. The amount of ***carbon dioxide*** oscillates with the seasons as deciduous trees and plants increase the amount they draw out of the air during their growing season, using it to build biomass. The amount of ***carbon dioxide*** in the atmosphere has increased in recent decades beyond the largest amount known in the past hundreds of thousands of years.

carbon equivalence A measure that compares the radiative impact of a given ***greenhouse gas*** to that of ***carbon dioxide***.

carbon sequestration The long-term storage, through natural processes or technology, of ***carbon*** in the terrestrial ***biosphere***, underground, or in the oceans. The aim is to reduce or slow the buildup of ***carbon dioxide*** in the atmosphere.

carbon trading An exchange mechanism for trading ***carbon*** emission rights as a means of meeting emission-reduction targets. Those who trade, buy, and sell a type of contract that allows them to emit, reduce emissions, or offset against emissions.

chaos theory The theory that minute inaccuracies in initial observations (such as temperatures or winds, in the climate change case), when placed into a forecast model will grow exponentially in time. It is also referred to as the *butterfly effect* after the idea that the impulse of a butterfly flapping its wings in Brazil could cascade and set off a tornado in Texas.

chlorofluorocarbons (CFCs) and related chemicals Synthetic chemicals that are odorless, nontoxic, nonflammable, and chemically inert; generally used for refrigerants, fire retardants, and aerosol sprays, they cause a seasonal decrease in the amount of protective ozone in the ***stratosphere***.

clean-development mechanism A device provided by the ***Kyoto Protocol*** to allow developed countries to earn emissions credits for investing in emissions-reducing projects located in developing countries.

climactic niches Area providing acceptable conditions for a species to persist and reproduce.

climate An average of ***weather*** conditions over an extended period of time (months, years, or centuries). The ***World Meteorological Organization*** uses a thirty-year period. Climate is characterized by temperature, precipitation, cloud cover, humidity, and wind patterns.

climate change A significant change from one climatic condition to another. The term often refers to the buildup of man-made gases in the ***atmosphere*** that trap the sun's heat, causing changes in global ***weather*** patterns. Some use the term interchangeably with ***global warming***; many scientists use the term to include natural changes in climate.

computer modeling and climate models The major control system for studying the complexity of the earth system. Because scientists cannot conduct classical experiments to study climate change, they use models of varying and increasing complexity. The ***Intergovernmental Panel on Climate Change*** defines a *climate model* as a numerical representation of the ***climate*** system that is based on the physical, chemical, and biological properties of its components, of their interactions, and of their feedback processes and that accounts for all or some of their known properties.

conference of the parties (COP) The supreme decision-making body of the United Nations ***Framework Convention on Climate Change (FCCC)***. It is composed of all of the nations that have ratified the agreement. Its first session was held in Berlin in 1995, and it has met a dozen times thereafter in The Hague, Milan, Montreal, and other cities.

consilience The coming together of inductions. It occurs when one class of facts coincides with an induction from a different class.

cryosphere The frozen water—snow, ice, and permafrost—on and beneath the surface of the earth. It includes the Greenland and Antarctic ice sheets and sea ice in the Arctic and Southern Oceans. It represents about 1 percent of the water on the planet.

deduction The process of drawing logical inferences from a set of premises.

ecosystem An interactive community of living organisms and their physical environment.

El Nino–Southern Oscillation An ***atmosphere*** and ocean phenomenon that channels year-to-year fluctuations in ocean temperatures over the tropical east Pacific into global fluctuations in ***climate***.

emissions trading A market-based system allowing those who can reduce emissions at a lower cost to sell their emission-reduction credits at a profit to others. Buyers use the credits to offset their own emissions (such as using fossil-fuel powered electricity generation) or emissions of the goods or services they consume (such as running a carbon-neutral business).

ensemble A large collection of models used to explore diverse possible outcomes.

equilibrium A stable state in a dynamic system in which changes are balanced by other changes.

error bars Used in statistics, these are the range of values estimated to contain the actual value area within a specified level of confidence uncertainty, usually 95 percent.

falsification The position in the study of scientific thinking that holds that a theory can never be proven true but it can be proven false.

flexibility mechanisms Devices for fostering implementation of the ***Kyoto Protocol***'s target for reducing ***greenhouse-gas*** emissions while limiting costs. These include ***joint implementation***, the ***clean-development mechanism***, and ***emissions trading***.

forcing Any imposed mechanism that forces ***climate*** to change. Natural forcing results from volcanic eruptions and solar variability; man-made or ***anthropogenic*** forcing comes from use of ***greenhouse gases*** and ***aerosols***.

Framework Convention on Climate Change (FCCC) An agreement signed in 1992 at the United Nations Conference on Environment and Development, which set out general parameters and principles for international efforts to address ***climate change***, with the aim of stabilizing ***greenhouse gas*** concentration in the ***atmosphere*** at a level that would prevent dangerous ***anthropogenic*** interference with the climate system. The agreement went into effect in 1994.

Global Climate Change Initiative The U.S. program, started in the Bush (George W.) administration, that aims to reduce the ratio of ***greenhouse gas*** emissions to economic output through domestic voluntary actions and continued research on ***climate change***.

Global Climate Coalition A group of major national environmental organizations whose goal is to influence the official U.S. response to ***climate change***.

global warming See ***climate change***.

global warming potential (GWP) An index that estimates the heat-trapping efficiency of ***greenhouse gases.*** It represents the ratio of energy reemitted to the earth's surface during a year for a given gas compared to that of the same mass of carbon dioxide.

greenhouse effect The warming of the planet because of the existence of the earth's ***atmosphere.*** The effect is similar to how a plant warms when it is encased in a house of glass or how a blanket traps body heat. It provides that the average surface temperature of the earth warms to 15°C (59°F). Greenhouse gases absorb thermal radiation emitted from the earth's surface and then reradiate this energy back to the surface of the earth—allowing temperatures to be significantly warmer than they would be in the absence of an atmosphere.

greenhouse gas Certain trace gases in the earth's ***atmosphere*** that selectively absorb and trap longer wavelengths (infrared radiation) of energy emitted by the earth and reemit them back to the earth's surface, allowing for a significant warming of the earth's surface and of the lower atmosphere. They account for less than 3 percent of the atmosphere and include ***carbon dioxide*** (CO_2), ***methane*** (CH_4), ***nitrous oxide*** (N_2O), ***halocarbons***, ***ozone*** (O_3), water vapor (H_2O), ***perfluorinated carbons (PFCs)***, and ***hydrofluorocarbons (HFCs).***

greenhouse gas efficiency A measure of the ability of gases to absorb thermal radiation emitted by the earth; it is a function of the molecular structure of the gas.

greenhouse-gas performance standard A maximum allowable rate of ***greenhouse-gas*** emissions from electricity generation. In California, the standard will be no lower than levels achieved by a new combined-cycle natural gas turbine. Effective for all utilities by mid-2007, electric utilities in California are prohibited from entering into a long-term financial commitment (ownership or a contract of five years or more) for often-used power plants that exceed the greenhouse-gas emission performance standard.

gross domestic product A measure of the economic activity that takes place within a country's borders during a period of time, calculated as consumption plus investment plus government spending plus exports minus imports.

halocarbons Potent ***greenhouse gases*** that do not exist in nature. They include ***hydrofluorocarbons*** and ***perfluorinated carbons.***

hydrofluorocarbons (HFCs) Alternatives to ***chlorofluorocarbons*** that were introduced in response to the ***Montreal Protocol.*** They do not destroy ozone, but they are a potent greenhouse gas, although with less ***global warming potential*** than CFCs.

hydrologic cycle The exchange of water among the atmosphere, land, and oceans through processes including precipitation and evaporation.

hypothetico-deductive model An approach often associated with the scientific method wherein scientists develop hypotheses and then test them.

induction The process of generalizing from specific examples.

integrated assessment An analysis of the advantages and disadvantages of adopting a particular action. Assessments involve literature reviews, quantification of losses associated with an action or failure to act, and cost-benefit analyses of effects on current generations against those on future generations.

Intergovernmental Panel on Climate Change (IPCC) The main international body established in 1988 by the ***World Meteorological Organization*** and the United Nations Environment Program to assess ***climate-change*** science and provide advice to the international community. The IPCC is an international group of scientists who summarize the current understanding of ***climate change*** and predict how ***climate*** may evolve.

joint implementation One of the ***flexibility mechanisms*** defined in article 6 of the ***Kyoto Protocol***. To meet their commitments, developed countries may exchange emission-reduction units based on projects aimed at reducing emissions at sources or increasing removals by **sinks** of ***greenhouse gases***.

Kyoto Protocol The 1997 agreement that strengthened the United Nations ***Framework Convention on Climate Change***. It committed developed countries to individual, legally binding targets to limit or reduce their ***greenhouse-gas*** emissions. Individual targets for reducing ***greenhouse-gas*** emissions for developed countries are listed in an Annex to the Kyoto Protocol. These add up to a total cut in greenhouse-gas emissions of at least 5 percent from 1990 levels in the commitment period 2008 to 2012. The Kyoto Protocol entered into force in February 2005, and as of late 2006, 166 states and regional organizations had ratified it.

lithosphere The earth's crust and solid portion of the mantle.

Little Ice Age A relatively minor ***climate*** variation in the fifteenth to the nineteenth centuries that was characterized by temperatures that were 2°C (3.6°F) cooler than today in much of Europe.

longwave or thermal radiation Energy emitted at longer ***wavelengths*** such as from the earth.

Medieval Warm Period A relatively minor ***climate*** variation in the tenth to fourteenth centuries that was characterized by unusually warm weather in Europe.

meeting of the parties (MOP) A meeting of the parties to the United Nations ***Framework Convention on Climate Change*** that occurred after the entry into force of the ***Kyoto Protocol.*** At Montreal in 2005, the first meeting of the parties ran parallel to the ***conference of the parties (COP)*** to the Convention.

methane (CH_4) A ***greenhouse gas*** that is odorless, colorless, and flammable. Among its sources is animal waste.

model calibration Adjustments in the parameters of a model in attempts to better reproduce past data.

Montreal Protocol on Substances That Deplete the Ozone Layer An international agreement designed to protect the ***ozone*** layer by phasing out the production of several substances believed to be responsible for ***ozone depletion.*** Known as ozone-depleting substances, they are linked to what is commonly known as the *ozone hole.*

National Oceanic and Atmospheric Administration (NOAA) A federal agency within the U.S. Department of Commerce that focuses on the condition of the oceans and the ***atmosphere.***

network Interconnected people or things that may be used to communicate information, allocate access to farming or fishing resources, or transport people, goods, or services.

Nitrous oxide (N_2O) A ***greenhouse gas*** with a ***global warming potential*** of about 310. Sources include the burning of biomass, soil-cultivation practices, fertilizer use, fossil-fuel combustion, and nitric acid production.

no-regrets policy A public policy that favors options that have negative net costs and that generate both direct and indirect benefits that are large enough to offset the costs of implementation.

oceanic conveyor belt Overall ocean circulation. The three-dimensional ocean movement that encompasses the planet is a theoretical route driven by wind and by the ***thermohaline*** circulation.

Organization of Economic Cooperation and Development (OECD) An international body composed of thirty countries that addresses social issues and provides statistics. It facilitates multilateral agreements and aid to promote good governance and further economic growth.

ozone Three oxygen atoms combined into a molecule, which forms a protective layer in the ***stratosphere*** that limits the amount of ***ultraviolet light*** reaching the surface of the earth to generally healthful levels.

ozone depletion The reduction of protective ***ozone*** in the earth's ***stratosphere***. Ozone depletion allows increased levels of ***ultraviolet light*** to reach the Earth's surface, increasing the risk of skin cancer. Ozone has been reduced through chemical reactions flowing from the release of ***chlorofluorocarbons*** and other ozone-depleting substances produced by humans.

paleoclimatology The study of past or ancient climates.

perfluorinated carbons (PFCs) Chemical compounds used in aluminum manufacturing, semiconductor manufacturing, and refrigeration. They do not harm stratospheric ***ozone***, but they are potent ***greenhouse gases***. PFCs are regulated by the ***Kyoto Protocol***.

polar amplification The tendency of large temperature changes to be much greater at the North or South Pole than the global average. This is thought to be caused by changes in the area covered by ice and snow, which absorb less energy from the sun than other surfaces.

policy entrepreneur Individuals with deep knowledge on a topic who can influence policy outcomes because their expertise is not seen to be politically motivated.

positive feedback A change in one element or variable that provides a change in a system.

proxy indicator Substitutes for what is observable in nature. For climate changes, proxy data are natural recording systems of past climate found in sediments, ice cores, tree rings, and corals.

salinity The salt content of a substance such as water or soil measured as the percentage of dissolved salts by weight. It is also expressed in parts per million (ppm). Ocean salinity is also measured in practical salinity units (PSU). The salinity of the ocean varies by depth and location but is generally about 3.5 percent salt (35,000 ppm or 35 PSU).

scientific method The process of conducting scientific inquiry. There is no universally accepted scientific method, but there is some consensus about the standards and criteria for judging what makes scientific information a reliable basis for action.

shortwave radiation Energy emitted at short ***wavelengths*** such as from the sun.

sink Any process that removes ***greenhouse gases*** or their precursors and ***aerosols*** from the ***atmosphere***.

smog Visibly polluted air, historically caused by sulfur dioxide from burning coal mixed with fog. Since the widespread use of automobiles, *smog* also means photochemical smog, created by chemical reactions of sunlight, emissions from mobile and stationary fossil-fuel combustion, and vapors from widely used chemicals such as gasoline, industrial solvents, and pesticides.

stratosphere The region of the ***atmosphere*** above the ***troposphere*** extending roughs from about 10 kilometers (or roughly 6 miles) to about 50 kilometers.

thermohaline Density-driven circulation in the ocean on a large scale. It is caused by differences in ***salinity*** and temperature.

toxic waste Discarded gaseous, liquid, or solid substances that can cause death or injury to humans or the environment. In the United States, acute toxicity is measured by the amount of the substance that is lethal to 50 percent of test rodents after fourteen days.

troposphere The part of the earth's ***atmosphere*** from the ground level up to at least 5 miles above the earth's surface. In some places the troposphere extends to 9 miles above the earth's surface.

ultraviolet light Energy with wavelengths slightly shorter than visible light.

water density Weight per volume of water, which increases with salt, cold, and pressure. If water becomes salty, it becomes more dense and sinks below fresher water. Likewise, cold water is denser than warm water and sinks below warmer water.

wavelength The distance between adjacent crests in the wave.

weather The atmospheric conditions for an individual event or time. The current limit of the forecast of weather extends to about ten days.

World Meteorological Organization (WMO) A United Nations agency concerned with the international collection of meteorological data.

Younger Dryas A period nearly thirteen thousand years ago when glacial conditions rapidly returned to the high latitudes of North America and temperatures in northern Europe dropped nearly 5°C (9°F).

Contributors

John T. Abatzoglou received his PhD in earth system science at the University of California, Irvine. His research interests are in atmospheric dynamics with an emphasis on large-scale teleconnection patterns including El Niño and the North Atlantic Oscillation.

Joseph F. C. DiMento is director of the Newkirk Center for Science and Society and professor of planning policy, design, and law and society at the University of California, Irvine. He is the author or editor of a number of books, including *Managing Environmental Change* (1976), *The Consistency Doctrine and the Limits of Planning* (1980), *Environmental Law and American Business: Dilemmas of Compliance* (1986), and *The Global Environment and International Law* (2003). He has contributed to these topics in the legal and popular press and is a regular observer of climate-change negotiations.

Pamela Doughman is an assistant professor of environmental studies at the University of Illinois, Springfield. She is also an energy specialist with the Renewable Energy Program at the California Energy Commission. Recent publications include "Water Cooperation in the U.S.-Mexico Border Region," in Ken Conca and Geoffrey D. Dabelko, eds., *Environmental Peacemaking* (2002), and "Discourse and Water in the U.S.-Mexico Border Region," in Joachim Blatter and Helen Ingram, eds., *Reflections on Water: New Approaches to Transboundary Conflicts and Cooperation* (2001).

Richard A. Matthew is an associate professor of international and environmental politics in the Schools of Social Ecology and Social Science at the University of California, Irvine, and director of the Center for Unconventional Security Affairs (www.cusa.uci.edu) and the Global Environmental Change and Human Security Research Office (www.gechs.uci.edu), both at UC Irvine. Recent books and coedited volumes include *Contested Grounds: Security and Conflict in the New Environmental Politics* (1999), *Dichotomy of Power: Nation versus State in World Politics* (2002), *Conserving the Peace: Resources, Livelihoods, and Security* (2002), *Reframing the Agenda: The Impact of NGO and Middle Power Cooperation in International Security Policy* (2003), and *Landmines and Human Security: International Politics and War's Hidden Legacy* (2004).

Stefano Nespor, lecturer at the Polytechnic University and an attorney in Milan, is a leading environmental, labor, and administrative lawyer in Europe and editor of *Rivista Giuridica dell'Ambiente*. He is the author of numerous books on the environment and law, including a recently published book with Alda L. De Cesaris (*Le Lunghe estati calde. Il cambiamento climatico e il protocollo di Kyoto*, 2004) in Italian for the nonscientist on climate change.

Naomi Oreskes is an associate professor in the department of history and the program in science studies at the University of California, San Diego. Her research focuses on the historical development of scientific knowledge, methods, and practices in the earth and environmental sciences. Her most recent book is *Plate Tectonics: An Insider's History of the Modern Theory of the Earth* (2001), which was cited by *Library Journal* as one of the best science and technology books of 2002.

Andrew C. Revkin has written about science and the environment for two decades, most recently as a reporter for the *New York Times*. He is the author of several books, including *The North Pole Was Here: Puzzles and Perils at the Top of the World* (2006) and *The Burning Season: The Murder of Chico Mendes and the Fight for the Amazon Rain Forest* (revised edition, 2004). He has received many honors, including the National Academies Science Communication Award, two journalism awards from the American Association for the Advancement of Science, a 2006 John Simon Guggenheim Fellowship, and an Investigative Reporters and Editors Award.

Index